Small-molecule Transcription Factor Inhibitors in Oncology

Drug Discovery Series

Titles in the Series:

1: Metabolism, Pharmacokinetics and Toxicity of Functional Groups
2: Emerging Drugs and Targets for Alzheimer's Disease; Volume 1
3: Emerging Drugs and Targets for Alzheimer's Disease; Volume 2
4: Accounts in Drug Discovery
5: New Frontiers in Chemical Biology
6: Animal Models for Neurodegenerative Disease
7: Neurodegeneration
8: G Protein-Coupled Receptors
9: Pharmaceutical Process Development
10: Extracellular and Intracellular Signaling
11: New Synthetic Technologies in Medicinal Chemistry
12: New Horizons in Predictive Toxicology
13: Drug Design Strategies: Quantitative Approaches
14: Neglected Diseases and Drug Discovery
15: Biomedical Imaging
16: Pharmaceutical Salts and Cocrystals
17: Polyamine Drug Discovery
18: Proteinases as Drug Targets
19: Kinase Drug Discovery
20: Drug Design Strategies: Computational Techniques and Applications
21: Designing Multi-Target Drugs
22: Nanostructured Biomaterials for Overcoming Biological Barriers
23: Physico-Chemical and Computational Approaches to Drug Discovery
24: Biomarkers for Traumatic Brain Injury
25: Drug Discovery from Natural Products
26: Anti-Inflammatory Drug Discovery
27: New Therapeutic Strategies for Type 2 Diabetes: Small Molecules
28: Drug Discovery for Psychiatric Disorders
29: Organic Chemistry of Drug Degradation
30: Computational Approaches to Nuclear Receptors
31: Traditional Chinese Medicine
32: Successful Strategies for the Discovery of Antiviral Drugs
33: Comprehensive Biomarker Discovery and Validation for Clinical Application

34: Emerging Drugs and Targets for Parkinson's Disease
35: Pain Therapeutics; Current and Future Treatment Paradigms
36: Biotherapeutics: Recent Developments using Chemical and Molecular Biology
37: Inhibitors of Molecular Chaperones as Therapeutic Agents
38: Orphan Drugs and Rare Diseases
39: Ion Channel Drug Discovery
40: Macrocycles in Drug Discovery
41: Human-based Systems for Translational Research
42: Venoms to Drugs: Venom as a Source for the Development of Human Therapeutics
43: Carbohydrates in Drug Design and Discovery
44: Drug Discovery for Schizophrenia
45: Cardiovascular and Metabolic Disease: Scientific Discoveries and New Therapies
46: Green Chemistry Strategies for Drug Discovery
47: Fragment-Based Drug Discovery
48: Epigenetics for Drug Discovery
49: New Horizons in Predictive Drug Metabolism and Pharmacokinetics
50: Privileged Scaffolds in Medicinal Chemistry: Design, Synthesis, Evaluation
51: Nanomedicines: Design, Delivery and Detection
52: Synthetic Methods in Drug Discovery: Volume 1
53: Synthetic Methods in Drug Discovery: Volume 2
54: Drug Transporters: Role and Importance in ADME and Drug Development
55: Drug Transporters: Recent Advances and Emerging Technologies
56: Allosterism in Drug Discovery
57: Anti-aging Drugs: From Basic Research to Clinical Practice
58: Antibiotic Drug Discovery: New Targets and Molecular Entities
59: Peptide-based Drug Discovery: Challenges and New Therapeutics
60: Drug Discovery for Leishmaniasis
61: Biophysical Techniques in Drug Discovery
62: Acute Brain Impairment Through Stroke: Drug Discovery and Translational Research
63: Theranostics and Image Guided Drug Delivery
64: Pharmaceutical Formulation: The Science and Technology of Dosage Forms
65: Small-molecule Transcription Factor Inhibitors in Oncology

How to obtain future titles on publication:
A standing order plan is available for this series. A standing order will bring delivery of each new volume immediately on publication.

For further information please contact:
Book Sales Department, Royal Society of Chemistry, Thomas Graham House, Science Park, Milton Road, Cambridge, CB4 0WF, UK
Telephone: +44 (0)1223 420066, Fax: +44 (0)1223 420247,
Email: booksales@rsc.org
Visit our website at www.rsc.org/books

Small-molecule Transcription Factor Inhibitors in Oncology

Edited by

Khondaker Miraz Rahman
King's College London, UK
Email: k.miraz.rahman@kcl.ac.uk

and

David E. Thurston
King's College London, UK
Email: david.thurston@kcl.ac.uk

ROYAL SOCIETY OF CHEMISTRY

Drug Discovery Series No. 65

Print ISBN: 978-1-78262-145-4
PDF ISBN: 978-1-78262-401-1
EPUB ISBN: 978-1-78801-527-1
Print ISSN: 2041-3203
Electronic ISSN: 2041-3211

A catalogue record for this book is available from the British Library

The Royal Society of Chemistry is a charity, registered in England and Wales, Number 207890, and a company incorporated in England by Royal Charter (Registered No. RC000524), registered office: Burlington House, Piccadilly, London W1J 0BA, UK, Telephone: +44 (0) 207 4378 6556.

For further information see our web site at www.rsc.org

Printed in the United Kingdom by CPI Group (UK) Ltd, Croydon, CR0 4YY, UK

Preface

Transcription factors (TFs) are regulatory macromolecules that induce profound and sustained effects in cells by interacting with, and modulating the expression of, genes responsible for critical cellular processes. They play a crucial role in tumourigenesis, and a number of oncogenic transcription factors such as AP-1 (Activator Protein 1), NF-κB (Nuclear Factor κB) and the STAT (Signal Transducer and Activator of Transcription) family (*e.g.*, STAT-3 and STAT-5) are over-activated in human malignancies. It has also been shown that the loss of activity of transcription factors can lead to phenotypic differences in *eukaryotic* organisms. Differences in the DNA sequences of TF-binding sites and related gene expression have been shown to vary considerably in individual organisms, although how these differences affect phenotypes is often unknown. In particular, there is significant interest in understanding how these changes, which influence both phenotype and cellular processes, affect the growth and evolution of tumour cells, and their response to chemotherapy. This book was conceived to inform and update the reader on the latest developments and thinking around small-molecule transcription factor inhibitors.

Cancer cells require increased levels of transcription for growth and survival compared with normal cells, and TF inhibitors can target the proteins that form the transcription machinery. *Eukaryotic* transcription mechanisms are complex, and small-molecule inhibitors can either modulate the signalling pathways that affect the activation of TFs, or prevent their association with the transcriptional machinery. Selective TF inhibitors are considered to be potentially less genotoxic to healthy cell populations within an organism compared with commonly used cancer therapies, such as chemotherapy or radiation. There have been attempts to regulate both general and gene-specific TFs in order to control gene expression. Inhibiting

Drug Discovery Series No. 65
Small-molecule Transcription Factor Inhibitors in Oncology
Edited by Khondaker Miraz Rahman and David E. Thurston

Published by the Royal Society of Chemistry, www.rsc.org

general TFs should provide a global effect, although this could also lead to global non-specific toxicity. Therefore, gene-specific TFs are a preferred therapeutic target, as modulation of the activity of a single gene has the potential to inhibit the survival of tumour cells addicted to over-expression of the particular gene being targeted, while potentially minimising global toxicities.

Small-molecules can modulate the activity of a single TF by directly interacting with different domains of the TF protein. They could interact either at the protein–DNA interface (PDI) and prevent association of the TF with its consensus DNA sequence, or could interact in the protein domains between components of the TF (*i.e.*, inhibiting a protein–protein interaction, or PPI), thus preventing activation of the TF. The latter is a particularly attractive approach for dimeric TFs such as the STATs, hypoxia-inducible factor (HIF) and NF-κB that require dimer formation to translocate to the nucleus before DNA interaction and gene activation occurs. The former approach of designing small-molecules to interact with the consensus DNA sequence of a particular TF thus preventing its association with DNA has the potential to be highly selective. However, designing DNA-interactive molecules with a high level of sequence-specificity has proved challenging, and there is concern that molecules with a lower specificity level could themselves produce longer-term genotoxic toxicities.

One of the key challenges of targeting the PDI interface is that not all DNA-binding domains of TFs are properly folded in the absence of their DNA target sequence. For example, the DNA-binding domain of members of the basic leucine zipper (bZip) family such as C/EBPα is largely unstructured in the absence of DNA, only forming the fully functional structure in the presence of its consensus DNA sequence. Similarly, the PPI domains of several TFs share homology (*e.g.*, the SH2 domain of STAT family members, and the Rel domain of NF-κB) which can be problematic when targeting gene-specific TFs. Another problem is that the PPI domain is often not "druggable" due to the lack of suitable cavities or pockets in the mainly large flat surfaces of most TF proteins. This lack of obvious binding pockets for small molecules makes it difficult to identify potential inhibitors. Also, according to the literature, many examples of small-molecule PPI inhibitors are characteristically non-drug-like (*e.g.*, having multiple aromatic rings, large and flat shape, high molecular weight, being outside of Lipinski's Rules *etc.*), which has deterred researchers during the past two decades. The small-molecule TF inhibitors that have been reported in the literature display a wide range of activity against different tumour cell lines. Molecules that target the PDI interface are generally significantly more active, with concentrations giving 50% of maximum inhibition (IC_{50}) values in the nanomolar to picomolar range, compared with inhibitors that target the PPI interface, which are typically only active at the micromolar level. However, recent advances in computational methodologies have allowed significant improvements to be made to the design of small-molecule inhibitors of both types, with improvements to both potency and specificity.

Overall, improvements to our understanding of TFs and their binding domains will be essential for the future discovery and development of more effective inhibitors. Also, from a precision medicine perspective, recent developments in the sequencing of DNA taken from the tumour cells of cancer patients indicate that, in the future, it may be possible to identify which TFs are being upregulated in a patient's tumour cells, so that the most relevant TF inhibitor can be administered.

This book begins with three chapters highlighting the contribution of synthetic methodologies, computational approaches and natural products to the discovery and development of transcription factor inhibitors. These are followed by a chapter on the use of pyrrolobenzodiazepines as transcription factor inhibitors, and chapters on the inhibition of NF-κB and STAT3. The final chapter discusses the transcriptional control of promotor G-quadruplexes.

Khondaker Miraz Rahman and David E. Thurston
King's College London

Contents

Drug Discovery Series No. 65
Small-molecule Transcription Factor Inhibitors in Oncology
Edited by Khondaker Miraz Rahman and David E. Thurston

Published by the Royal Society of Chemistry, www.rsc.org

CHAPTER 1

Synthetic Approaches and Challenges to Transcription Factor Inhibitors

GEORGE PROCOPIOU[a] AND PANAYIOTIS A. PROCOPIOU*[b]

[a] Institute of Pharmaceutical Science, Britannia House, Kings College London, 7 Trinity Street, London SE1 1DB, UK; [b] GSK Medicines Research Centre, Medicinal Chemistry, FLI DPU, Respiratory Therapeutic Area, GlaxoSmithKline, Gunnels Wood Road, Stevenage SG1 2NY, UK
*Email: pan@procopiou.net

1.1 Introduction

Over the past two decades, transcription factors have attracted considerable research interest due to their potential use as chemotherapy agents for the treatment of various human cancers. Transcription factors are proteins that in cancers are known to become overactive and instigate the excessive transcription of genes, the products of which are required for the growth of tumors and in metastasis.[1]

In order to inhibit the processes of a particular transcription factor, the specific disruption of the interaction between two proteins or between DNA and a protein is needed. Until relatively recently, it has generally proven difficult to identify small molecules that effectively inhibit such interactions, one factor being the lack of distinct and specific binding sites.[2] Furthermore, small molecules aimed at potentially inhibiting transcription factors must contend with a large free energy of association between protein–protein and protein–DNA interfaces. There are considered to be four

Drug Discovery Series No. 65
Small-molecule Transcription Factor Inhibitors in Oncology
Edited by Khondaker Miraz Rahman and David E. Thurston

Published by the Royal Society of Chemistry, www.rsc.org

potential target areas for small-molecule transcription factor inhibitors: the DNA binding domain, the dimerization domain (of dimeric transcription factors), the coactivators/mediators of transcription, and the cognate DNA sequence of specific transcription factors. Compounds belonging to many different chemical classes have been reported to inhibit transcription factors, including polyamides, peptidomimetics, nucleosides, alkaloids, natural products and even some complexes of transition metals.[1b]

Clearly, an important hurdle in the progression of any promising hit or lead to a drug candidate will be its synthetic viability for the efficient and timely delivery of significant quantities of material to be used either for further testing or for the lead-optimization process of analogues.[3]

Hence, this chapter will focus on some of the interesting synthetic chemistry that has been used to access these targets, especially where significant challenges remain and non-trivial or atypical reactions are needed. By no means exhaustive, the small-molecule compounds covered in this section will be mostly limited to those that have either shown promise due to their higher potency *in vitro* and/or have already advanced to a late stage in the drug discovery process. Moreover, compounds that are not conventionally considered 'drug-like' and are therefore unlikely to be developed as therapeutic agents will not be considered, unless they constitute new hits for further investigation. Lastly, since the actual identification of new transcription factor inhibitors is itself a challenge, innovative approaches towards these ends are also described.

1.2 Cyclin-Dependent Kinase Inhibitors—Flavopiridol

Flavopiridol/Alvocidib (**1**) [2-(2-chlorophenyl)-5,7-dihydroxy-8-[(3*S*,4*R*)-3-hydroxy-1-methyl-4-piperidinyl-4*H*-1-benzopyran-4-one] [146426-40-6] (Figure 1.1) is a small molecule (molecular weight 401.8) synthetic flavone derivative that has been shown to possess antitumor activity against human lung and breast carcinoma. Flavopiridol was being developed by Aventis and it has progressed to multiple Phase IIb clinical trials as an anticancer agent.

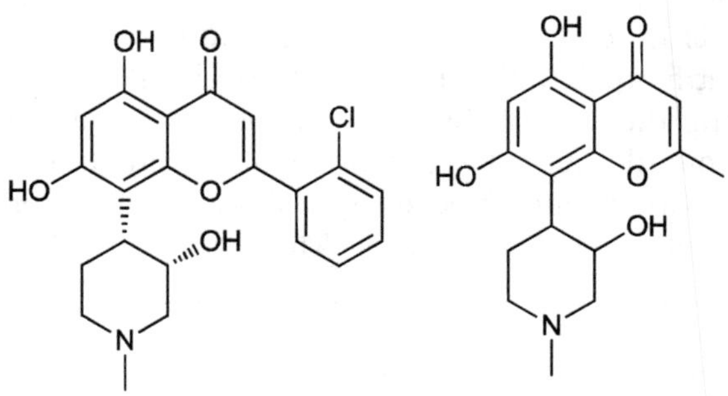

1 **2**

Figure 1.1 Structures of Flavopiridol (**1**) and a related natural product (**2**).

It is a potent and selective inhibitor of the cyclin-dependent kinases (CDKs), which are serine and threonine protein kinases that drive cell cycle and cell proliferation. It was subsequently found that Flavopiridol disrupts signal transducer and activator of transcription 3 (STAT3)–DNA interactions, and attenuates STAT3-directed transcription.

Flavopiridol was chosen as a representative example from this class of inhibitors as it progressed into development and attracted a lot of attention. It is related to the natural product (+)-*cis*-5,7-dihydroxy-2-methyl-8-[4-(3-hydroxy-1-methyl)piperidinyl]-4*H*-1-benzopyran-4-one (**2**), isolated from the stem-bark of the Indian plant *Dysoxylum binectariferum*, which differs from **1** by the substituent at the 2 position of the pyranone ring (methyl instead of 2-chlorophenyl). The natural product **2** was isolated with 0.9% yield of the dry weight of the plant material.[4]

A total synthesis of **2** was undertaken as part of its structure elucidation and formed the basis of subsequent syntheses of **1** and its various analogues. Reaction of 1,3,5-trimethoxybenzene with *N*-methyl-4-piperidone in a mixture of glacial acetic acid and hydrochloric acid at 95 °C gave **3** with 80% yield (Scheme 1.1). *In situ* hydroboration of the olefin **3** with borane generated from sodium borohydride and boron trifluoride etherate, followed by oxidation with alkaline hydrogen peroxide gave the *trans* (±) **4** at 70%. Initial attempts to invert the configuration of the hydroxy group were unsuccessful, however Swern oxidation to ketone **5** (76%), followed by sodium borohydride reduction gave a mixture of *cis* and *trans* alcohols in a 7 : 3 ratio. Crystallization of the hydrochloride salt of this mixture of alcohols gave the racemic *cis*-alcohol **6** at 66% yield. The latter was resolved using (−)-dibenzoyl-D-tartaric acid and the resulting solid was recrystallized five times from methanol and diisopropyl ether. The crystalline tartrate was basified to give (+)-**6** (20%). The filtrate from the crystallizations was basified and the free base was treated with (+)-dibenzoyl-L-tartaric acid and recrystallized five times from methanol diisopropyl ether to give after basification the free base of (−)-**6** (9%). The (−) enantiomer of **6** was treated with boron trifluoride etherate in acetic anhydride to give selectively the mono-demethylation product, which underwent a Fries rearrangement to the acetophenone derivative, and then alkaline hydrolysis gave the diol **7** at 73%. Condensation with ethyl acetate in the presence of sodium gave the benzopyranone **8**, which was finally demethylated with pyridinium hydrochloride in the presence of quinoline at 180 °C to give **2** at 14% (for the racemate, the yield for the natural product was not reported).

Synthesis of flavopiridol **1** followed the route outlined above for the natural product **2** and is reported in the patent literature.[5,6] The synthesis started with diol **7**, which was reacted with 2-chlorobenzoyl chloride in pyridine, followed by cyclisation (NaH in THF) to give the dimethoxybenzopyranone **9**. The latter was demethylated using the same conditions as before to give **1** (Scheme 1.2). This synthetic route was used by the Mitotix group[7] to explore structure–activity relationships (SAR) of the 3-hydroxy-1-methylpiperidinyl ring of flavopiridol, and the BMS group[8] to investigate thio- and oxo-flavopiridol analogues. The BMS group improved the synthesis by conducting

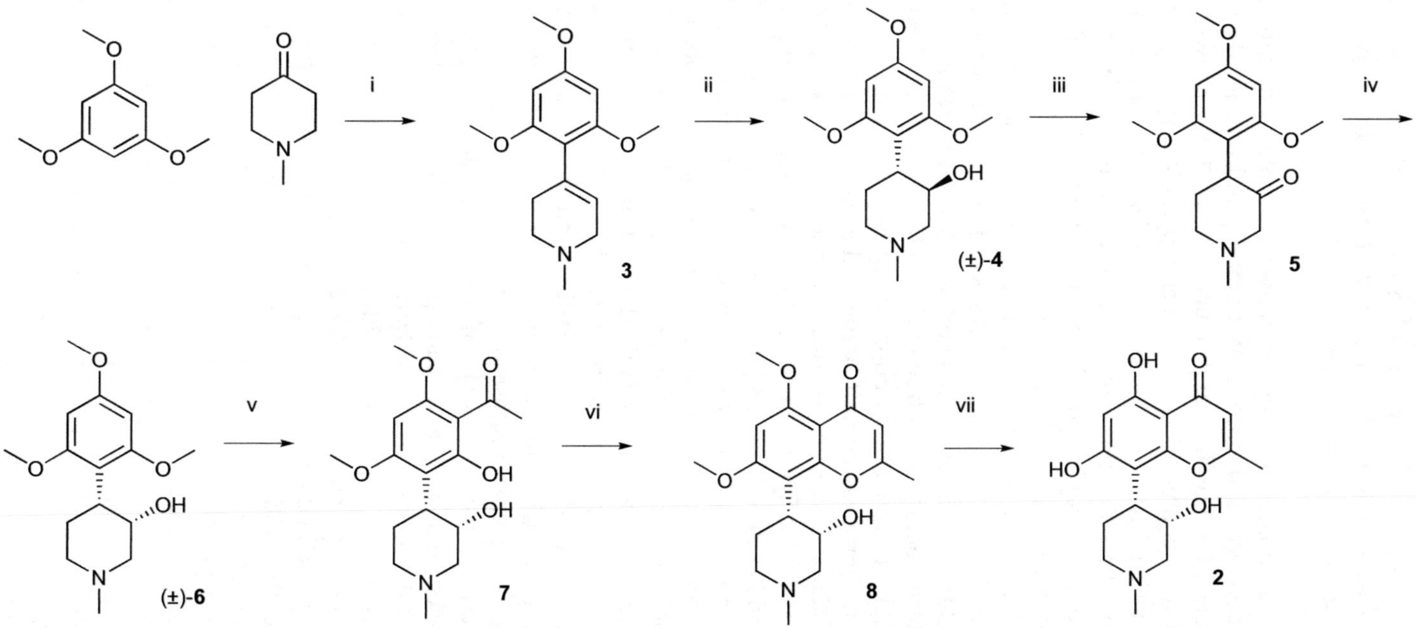

Scheme 1.1 Reagents and conditions: (i) HCl, AcOH, 95 °C, 3 h, 80%; (ii) (a) $NaBH_4$, $BF_3 \cdot OEt_2$, (b) 30% H_2O_2, NaOH, 70%; (iii) Swern oxidation, 76%; (iv) (a) $NaBH_4$, EtOH, 66%; (b) resolution with (−)-dibenzoyl D-tartaric acid gave (+)-**6** (20%) and (−)-**6** (9%); (v) (a) $BF_3 \cdot OEt_2$, Ac_2O, (b) NaOH, 73%, (vi) EtOAc, Na, reflux, 3 h; (vii) pyridinium hydrochloride, quinoline 180 °C, 2 h, 14%.

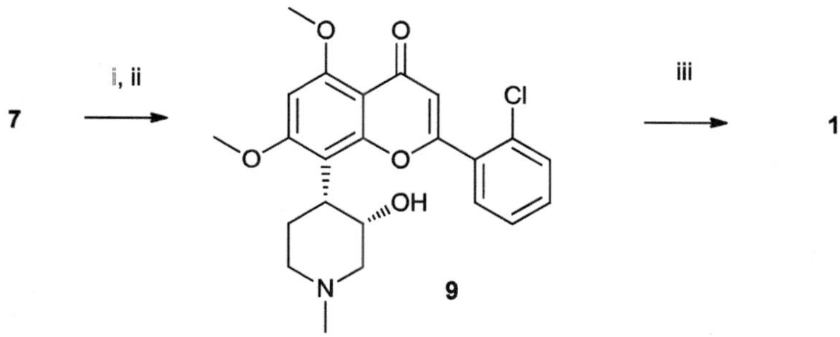

Scheme 1.2 Reagents and conditions: (i) 2-chlorobenzoyl chloride, pyridine; (ii) (a) NaH, THF; (b) HCl; (iii) pyridinium hydrochloride, 180 °C.

a dynamic resolution of the ketone **5** using dibenzoyl-D-tartaric acid in methanol in which the desired *R*-enantiomer was insoluble, whereas the *S*-enantiomer underwent a facile epimerization. Ketone (*R*)-**5** was obtained at 76% yield at a 100 g scale using this improved method. Diisobutylaluminium hydride (DIBAL-H) reduction of (*R*)-**5** gave the enantiopure *cis*-alcohol (3*S*,4*R*) **6** (56%) and the *trans*-alcohol **4** (13%) after flash chromatography on silica, followed by separation on a Chiralcel AD column.

Finally, a stereocontrolled synthesis to flavopiridol analogues substituted in the piperidine ring was published by the Aventis group and is summarized in Scheme 1.3.[9] The synthesis commenced with Wadsworth–Emmons olefination of 2,4,6-trimethoxybenzaldehyde with trimethyl phosphonoacetate, followed by DIBAL-H reduction of the resulting ester to give the allylic alcohol **10** (88%). Claisen–Johnson rearrangement of **10** followed by base hydrolysis of the resulting ester gave acid **11** (51%). Iodolactonisation and iodide displacement with sodium azide gave the lactone **12** (70%), which was enolized with lithium bis(trimethylsilyl)amide (LiHMDS) and methylated with iodomethane at −70 °C to generate the *trans-trans* lactone **13** (60%). The azide group was reduced by hydrogenation over 10% Pd/C at elevated pressure (100%) and then treatment with sodium methoxide in methanol gave the lactam **14** at 75% yield. The hydroxy group was protected as the *tert*-butyldimethylsilyl (TBDMS) ether, the lactam was reduced with lithium aluminium hydride (LAH) and the piperidine nitrogen was reductively methylated with simultaneous loss of the TBDMS group to give the tertiary amine **15** at 76% yield for the three steps. The remaining steps followed the flavopiridol chemistry to form first the acetophenone **16** *via* selective demethylation–Fries rearrangement (30–50%), condensation of **16** with methyl 2-chlorobenzoate, followed by cyclodehydration to the pyranone **17** (30%), and final demethylation using pyridinium hydrochloride gave the flavopiridol analogue **18** (the yield for the last step was not reported).

The routes outlined above are far too long, linear and not stereospecific, relying on resolution using dibenzoyl-D-tartaric acid. There are low-yielding steps at the end of the synthetic route, such as the last demethylation step (14%), making the synthetic route in need of improvement. Despite the

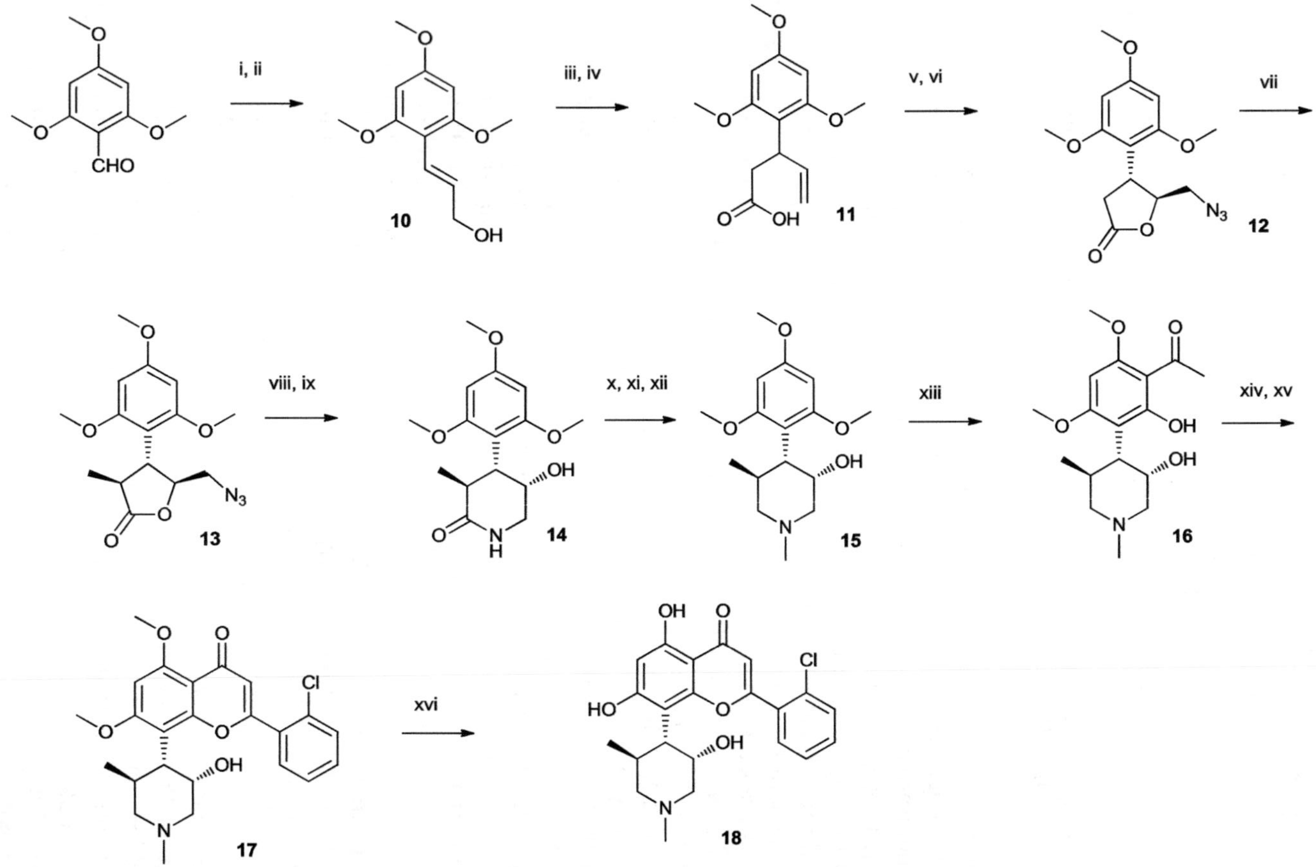

Scheme 1.3 Reagents and conditions: (i) $(MeO)_2POCH_2CO_2Me$, THF, LiHMDS, 0 °C, 90%; (ii) DIBAL-H, PhMe, 20 °C, 98%; (iii) $MeC(OEt)_3$, pivalic acid, *o*-xylene, 135 °C, 60%; (iv) NaOH, MeOH, 50 °C, 85%; (v) I_2, MeCN, 0 °C, 80%; (vi) NaN_3, DMF, 100 °C, 88%; (vii) LiHMDS, THF, −70 °C, MeI, 60%; (viii) H_2 (60 psi), EtOH, 10% Pd/C, 100%; (ix) NaOMe, MeOH, 65 °C, 75%; (x) TBDMS-Cl, imidazole, DMAP, DMF, 95%; (xi) LAH, THF, 65 °C, 90%; (xii) aqueous CH_2O, $NaBH_3CN$, pH 5, 90%; (xiii) Ac_2O, $BF_3 \cdot OEt_2$, NaOH, 30–50%; (xiv) DMF, 95%, NaH, methyl 2-chlorobenzoate, 30%; (xv) 4M HCl, dioxane; (xvi) pyridinium hydrochloride, quinoline, 160 °C.

improvement in the synthesis using the dynamic resolution of ketone **5** the next step (DIBAL-H reduction) required extensive and expensive chromatographic separation on a Chiralcel AD column. The route outlined in Scheme 1.3 was also very long (16 steps), provided racemic product and suffered from a number of low-yield steps.

1.3 Inhibitors of STAT3 and STAT5

Signal transducer and activator of transcription proteins (STATs) are cytoplasmic transcription factors that transmit extracellular signals from receptors on the plasma membrane to the nucleus where they bind to various promoters and initiate gene transcription. They play a critical role in the growth and survival of human tumor cells. Inhibition of STAT3 signaling induces growth arrest and apoptosis. STAT activation is dependent on tyrosine phosphorylation. Turkson and colleagues,[10,11] and McMurray and colleagues[12–17] have developed several drug-like peptidomimetic inhibitors of STAT3 based on a dimerization-disrupting phosphorylated peptide sequence derived from the SH2 domain-binding region of STAT3. Representative examples **19–25** are shown in Figure 1.2. These peptidomimetics are quite potent *in vitro* and compound **22** for example has a concentration giving 50% of maximum inhibition (IC_{50}) of 69 nM. However, these compounds lack whole-cell activity due to their high polarity and dianionic nature, which restrict their cell-membrane permeability. Furthermore, as they are tyrosinephosphate derivatives they are labile towards phosphatases. For these reasons McMurray and colleagues replaced the phosphate group with the phosphatase-stable phosphonodifluoromethyl group (compound **23**). Additionally, they used the pivaloyloxymethyl pro-drug approach to increase the lipophilicity of these compounds and increase their cell-permeability. It was observed that the dipivaloyloxymethyl ester pro-drugs exhibited dose-dependent inhibition of STAT3 in cell assays, whereas the mono ester pro-drugs were inactive at 25 μM. Compound **23** was selective for STAT3 over STAT1, STAT5, Src and p85 of phosphoinositide 3-kinase (PI3K). The pro-drug was not cytotoxic to a range of tumor cells at concentrations that completely inhibited STAT3 phosphorylation.

The synthetic challenge for the simpler tyrosinephosphate analogues was in the synthesis of the non-proteinogenic amino acids. Standard peptide chemistry was used to assemble the phosphopeptidomimetic inhibitors, however, these will not be discussed further. The synthesis of double pro-drug **23**, which includes the incorporation of the tyrosine isostere, is more challenging than that of the other inhibitors of this class, as described below. The synthesis of the difluoromethylphosphonocinnamate **26** is shown in Scheme 1.4, and it commenced with the olefination of 4-iodoacetophenone (**27**) with *tert*-butyl diethylphosphonoacetate in the presence of lithium *tert*-butoxide to give **28** (79%).[16] Diethyl bromodifluoromethylphosphonate was reacted with cadmium powder and the resulting organocadmium reagent was added to **28** to give the aryldifluoromethylphosphonate **29** with 68% yield. The

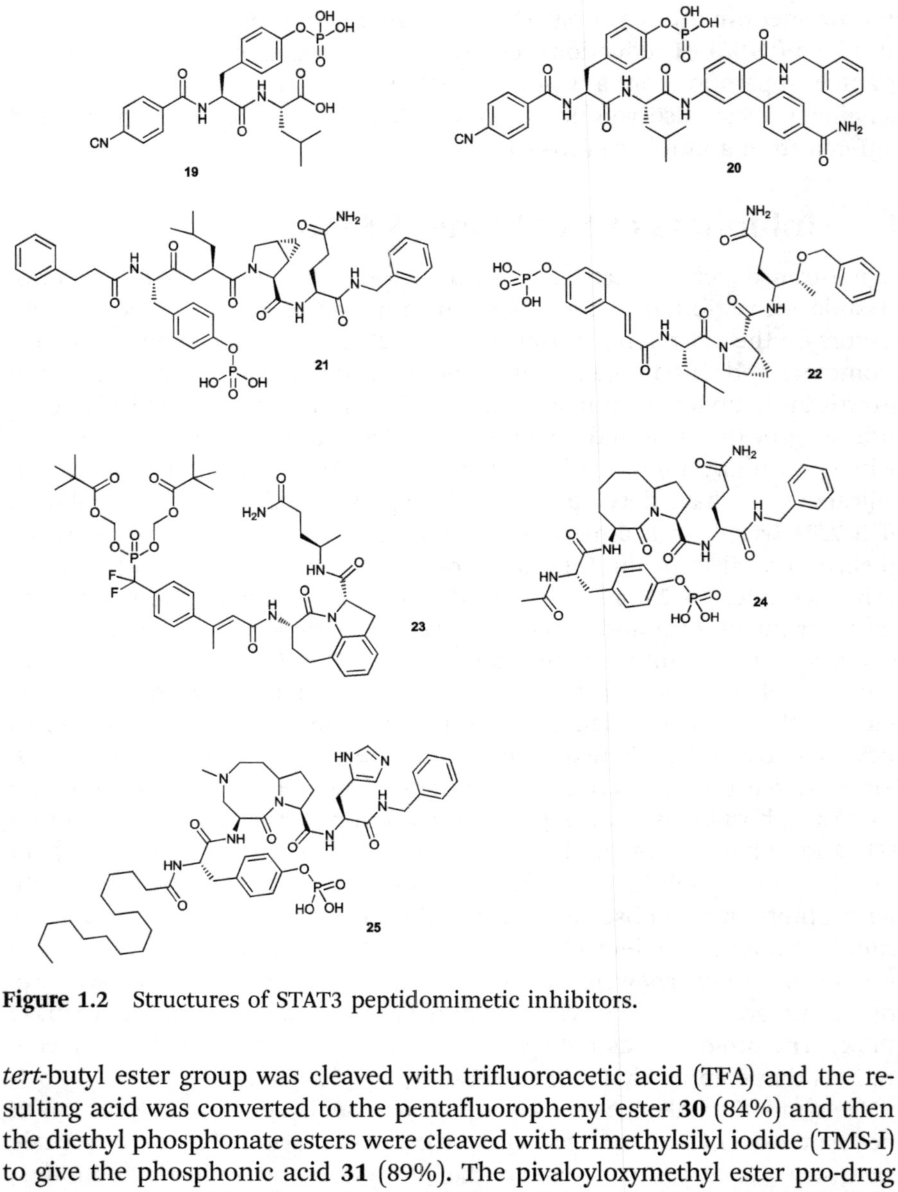

Figure 1.2 Structures of STAT3 peptidomimetic inhibitors.

tert-butyl ester group was cleaved with trifluoroacetic acid (TFA) and the resulting acid was converted to the pentafluorophenyl ester **30** (84%) and then the diethyl phosphonate esters were cleaved with trimethylsilyl iodide (TMS-I) to give the phosphonic acid **31** (89%). The pivaloyloxymethyl ester pro-drug moiety was assembled from commercially available pivaloyloxymethyl iodide using silver nitrate to produce **26** at 64% yield after chromatography.

The core azepino-indoline amino acid **32** [(2*S*,5*S*)-5-(9-fluorenyloxycarbonyl)-amino-1,2,4,5,6,7-hexahydro-4-oxoazepino[3,2,1-hi]indole-carboxylic acid] was synthesized in five steps[14] using the procedure first reported by De Lombaert *et al.* for use in angiotensin-converting enzyme inhibitors and is outlined in Scheme 1.5.[18] Reaction of methyl (*S*)-indoline-2-carboxylate (**33**) with (*S*)-*N*-trifluoroacetyl aspartic anhydride (**34**) in ethyl acetate gave a mixture of the two regioisomeric products (**35** and **36**) in quantitative yield and at a ratio of

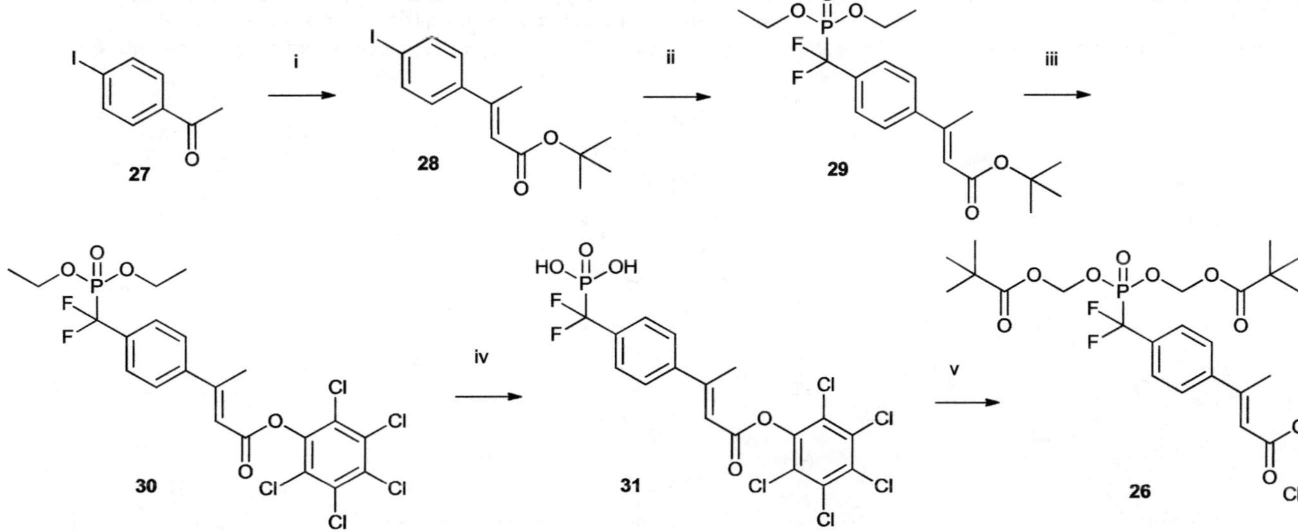

Scheme 1.4 Reagents and conditions: (i) $(EtO)_2POCH_2CO_2{}^tBu$, 79%; (ii) $BrCdCF_2PO(OEt)_2$, CuCl, 68%; (iii) (a) TFA, (b) C_6Cl_5OH, DCC, 84%; (iv) TMS-I, 89%; (v) (a) NaOH, (b) $AgNO_3$, (c) pivaloyloxymethyl iodide.[16]
Reprinted with permission from P. K. Mandal, F. Gao, Z. Lu, Z. Ren, R. Ramesh, J. S. Birtwistle, K. K. Kaluarachchi, X. Chen, R. C. Bast Jr, W. S. Liao and J. S. McMurray, *J. Med. Chem.*, 2011, **54**, 3549. Copyright 2011, American Chemical Society.

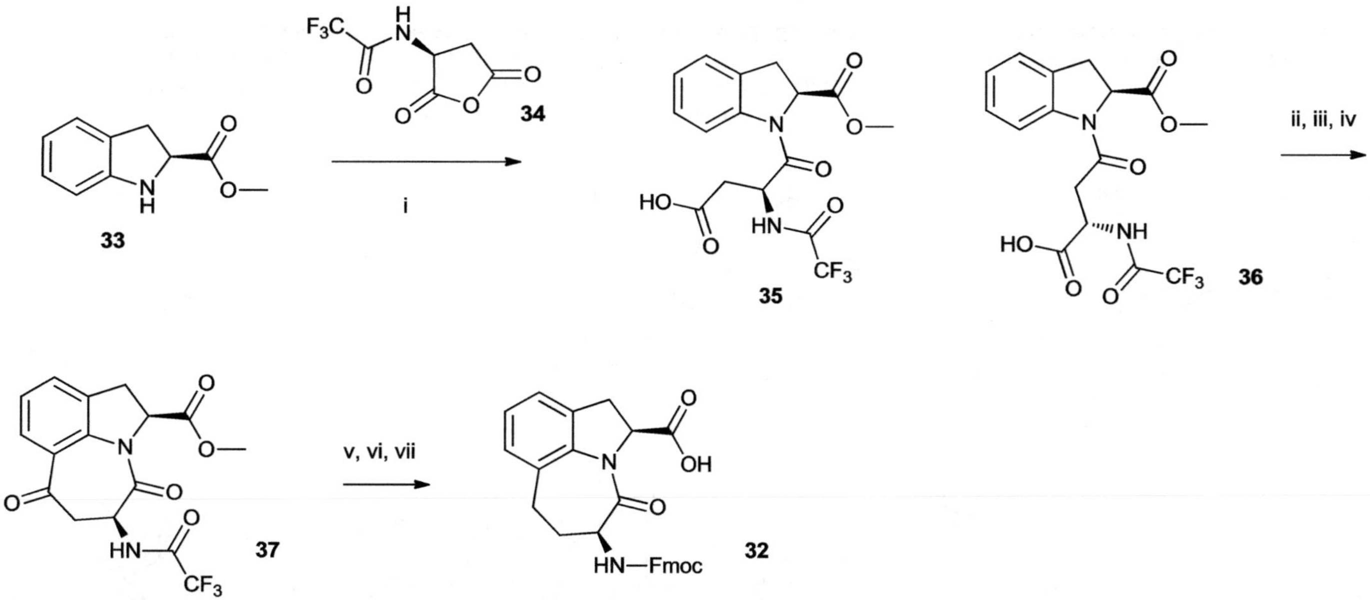

Scheme 1.5 Reagents and conditions: (i) EtOAc, 2 h; (ii) oxalyl chloride, DMF (catalyst), $ClCH_2CH_2Cl$; (iii) $AlCl_3$, 50C; (iv) DBU (cat.), dioxane; (v) H_2, Pd/C, AcOH; (vi) LiOH, THF; (vii) Fmoc-Osu.
Reprinted with permission from P. K. Mandal, D. Limbrick, D. R. Coleman, G. A. Dyer, Z. Ren, J. S. Birtwistle, C. Xiong, X. Chen, J. M. Briggs and J. S. McMurray, *J. Med. Chem.*, 2009, **52**, 2429. Copyright 2009, American Chemical Society.

87 : 13 respectively. The mixture was taken directly to the next stage, which involved conversion to the acid chloride, followed by Friedel–Crafts acylation/cyclisation to yield a regioisomeric mixture of the tricyclic azepine derivative. The mixture was equilibrated with a catalytic amount of DBU to give the (2*S*, 5*S*) diastereoisomer **37** as the major product (92 : 8), which was isolated in a pure form after fractional crystallization. The ketone group of **37** was hydrogenated over Pd/C, the amide and ester groups hydrolyzed with LiOH, and the amine group capped with 9-fluorenylmethyloxycarbonyl (Fmoc) to give **32**.

The synthesis of **23** was completed using standard peptide coupling chemistry on Rink resin attaching first Fmoc-(*R*)-4-aminopentanoic acid, followed by **32** using three-fold excess of the Fmoc-amino acids, (benzotriazol-1-yl-oxy)tripyrrolidinophosphonium hexafluorophosphate (PyBOP), and 1-hydroxybenzotriazole (HOBt) with six-fold excesses of *N*,*N*-diisopropylethylamine (DIPEA) in dimethylformamide–dichloromethane (DMF–DCM). The Fmoc group was removed with piperidine in DMF, the product was cleaved from the resin using TFA and purified by reverse-phase HPLC. The purified product was then coupled with **26** in the presence of HOBt and DIPEA in *N*-methylpyrrolidone (NMP) to give crude **23**, which required further purification by reverse phase HPLC.[16] The challenges associated with this important new area of research are the high number of steps, the requirement for excess Fmoc-amino acids for the couplings, the requirement for reverse-phase purifications, the use of organocadmium reagents and the use of silver nitrate. The synthetic route however is highly convergent.

The cancer-relevant transcription factors STAT5a and STAT5b are particularly challenging targets for identifying selective small-molecule inhibitors, as they are highly homologous proteins (over 90% amino acid sequence identity) and with similar binding preferences. Very recently the very first small molecule nanomolar inhibitor of STAT5b, named Stafib-1 (**38**, Figure 1.3) with more than 50-fold selectivity over STAT5a has been reported by Berg and colleagues.[19] The pivaloyloxymethyl pro-drug ester of **38**, named Pomstafib-1 (**39**), inhibited tyrosine phosphorylation of a STAT5b–green fluorescent protein (GFP) fusion protein with high selectivity in human leukemia cells, demonstrating that the pro-drug had cell-permeability and was hydrolyzed by intracellular esterases, liberating **38**. The chemistry to synthesize both the parent active molecule and the prodrug will not be outlined here as it is straightforward, and has some similarities with the chemistry already presented above. It is worth noting that **38** is a prototype tool compound for further investigations in mammalian cells. Compound **38** is vulnerable to metabolic clearance, such as hydrolysis of the catechol phosphates by phosphatases, oxidation of the benzylic amine group, hydrolysis of the glycolamide and hydrolysis of the anilide to generate toxic aniline.

1.4 Inhibitors of c-Myc

c-Myc is a basic helix–loop–helix leucine-zipper transcription factor that is important in fundamental cellular processes, which include cell cycle

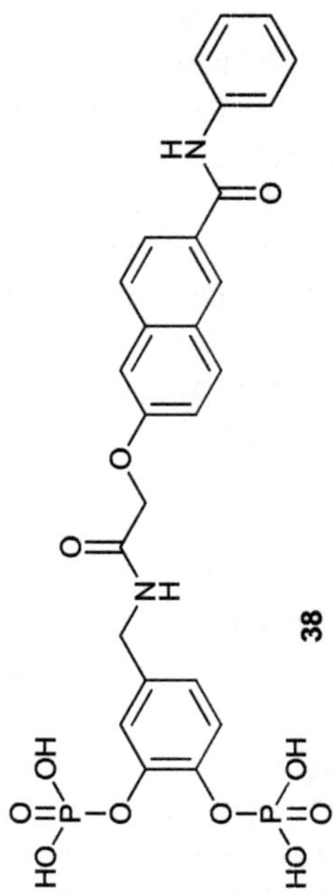

38

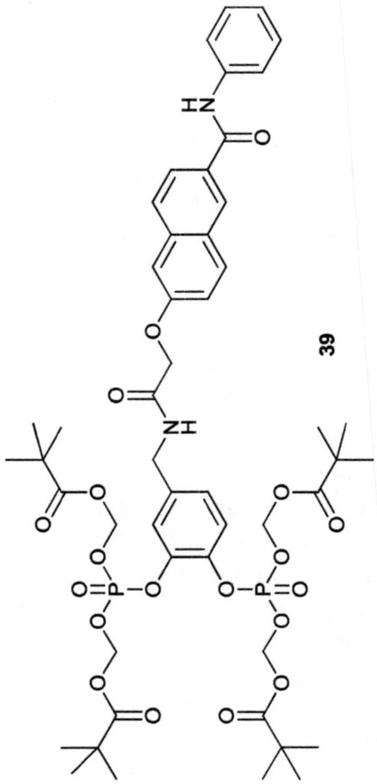

39

Figure 1.3 Structure of STAT5b inhibitor Stafib-1 (**38**) and the pro-drug Pomstafib-1 (**39**).

progression, growth, oncogenic transformation and apoptosis.[20] c-Myc is pathologically activated and upregulated in human cancers.[21–23] As in the case of the STATs, disturbing the dimerization of a crucial transcription factor could be a promising strategy to exert influence on gene expression and hold up the dimerization process and could potentially lead to novel anti-cancer therapies where elevated levels of c-Myc are observed. The activation is driven by heterodimers formed from c-Myc with its partner protein Max. The first interaction for which a small-molecule antagonist was identified was for Myc–Max.[24] Several antagonists were discovered in this study with the best antagonist being the isoindoline **40** (50 μM). The rational design of c-Myc antagonists is far more challenging than that for STAT3 antagonists. Dimerization of STAT3 is initiated by the phosphorylation of a conserved tyrosine residue (Tyr705) C-terminal to the Src-homology 2 (SH2) domain, which leads to dimerization through reciprocal SH2-phosphotyrosine (pTyr) interactions. In the case of c-Myc–Max there is no obvious binding site and c-Myc exhibits no recognizable tertiary or secondary structure.[25] For this reason, the majority of c-Myc antagonists reported so far have been identified from high-throughput screening of diverse chemical libraries. Representative c-Myc antagonists are shown in Figure 1.4.[24,26–30]

The antagonists **40–42** are all high-molecular-weight (893, 680 and 721 respectively) with activities greater than 17 μM. Attempts were made to obtain smaller inhibitors and several were identified with molecular weights of less than 500, for example **43** and **44**, however their activities were between 30 and 50 μM.[28–30] All of the reported c-Myc–Max antagonists represent initial leads and are currently far from being developed as potential anti-cancer agents. As the above compounds originated from chemical libraries, we outline in Scheme 1.6 the synthetic route used to make the chemical library from which compound **42** was identified.

Thus, treatment of **45** with LiF generated the azomethine ylide, which underwent a [3 + 2] cycloaddition with dimethyl fumarate to provide the pyrrolidine dicarboxylate **46**. The latter was deprotected by hydrogenolysis over $Pd(OH)_2$ to give **47**, and then the pyrrolidine amine group was acylated with six carboxylic acids to give six amides **48**. The methyl esters were hydrolyzed to give six dicarboxylic acids **49**, which were reacted with one equivalent of each of 20 amines to give 120 amides **50**. From the 120 amidocarboxylic acids **50** tripeptides **51** were obtained to give a library of 240 compounds from which **42** was identified.

Very recently the Fletcher group has published a novel approach based on synthetic α-helix mimetics that inhibit a range of helix-mediated protein–protein interactions in which only one of the protein partners uses an α-helical recognition domain. They proposed that rationally designed α-helix mimetics might disrupt the coiled coil of the c-Myc–Max heterodimer, thus providing a novel therapeutic strategy to inhibit the oncogenic function of c-Myc.[31] The small-molecule compound **52** (MW 514) was identified as the most active inhibitor, with an *in vitro* IC_{50} value of 5.6 μM. Furthermore, under conditions where **52** inhibited cell growth, c-Myc–Max heterodimers remained intact. However, **52** has a short half-life due to metabolism of the nitro group

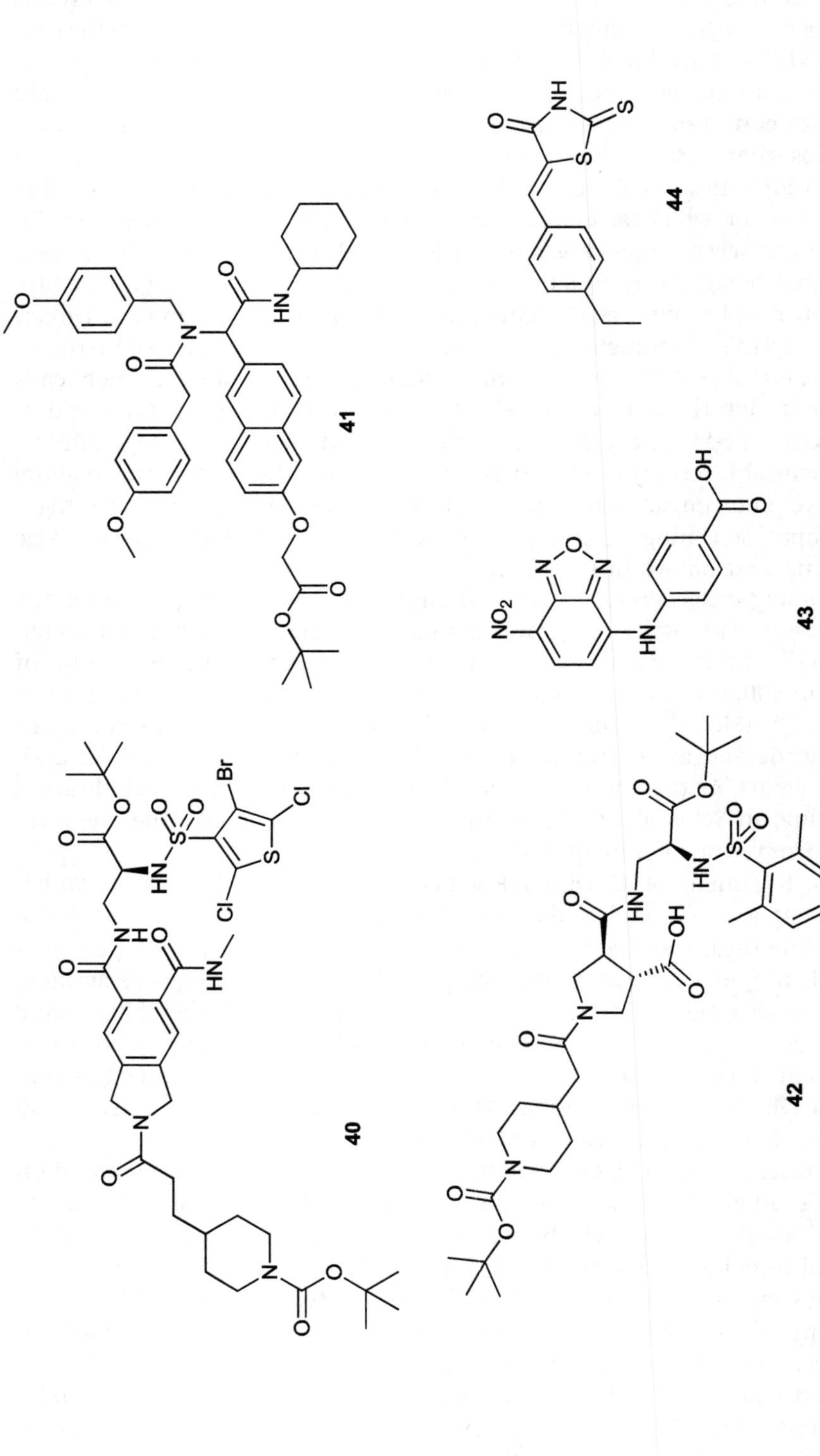

Figure 1.4 Small-molecule antagonists of c-Myc.

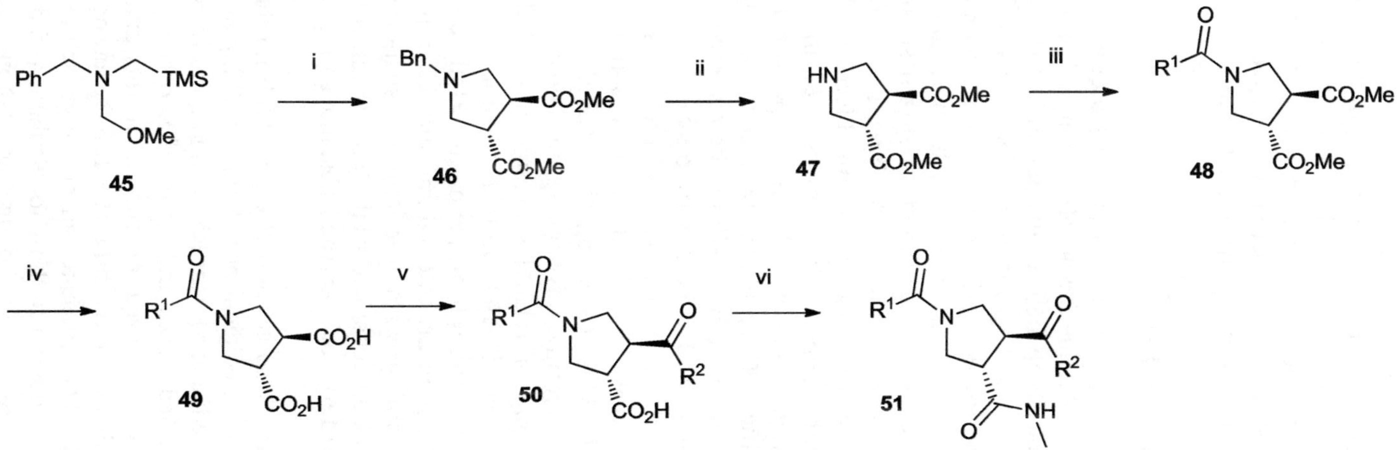

Scheme 1.6 Reagents and conditions: (i) LiF, dimethyl fumarate, MeCN, 96%; (ii) H_2, $Pd(OH)_2$; (iii) R^1CO_2H, EDCI; (iv) hydrolysis; (v) 20 amines, EDCI; (vi) $MeNH_2$, EDCI.

to the corresponding hydroxylamine and aniline, making it a hit compound for further investigation. Compound **52** was synthesized by the short and convergent route shown in Scheme 1.7 starting from 3-fluoro-4-nitrobenzoic acid (**53**). The ether group was introduced by aromatic nucleophilic substitution with alkoxide to provide **54**, which was then converted with oxalyl chloride to the acid chloride **55**. The latter was reacted with aniline **56** to produce anilide **57**, which was finally hydrolyzed to the required acid **52**.

1.5 Inhibitors of HIF-1

Another important transcription factor that plays a crucial role in various diseases is the hypoxia-inducible factor (HIF) group of proteins. HIF-1 is a heterodimeric, basic helix–loop–helix transcription factor, comprised of two subunits: HIF-1α and HIF-1β. HIF-1α is more important as it activates a large number of genes in the angiogenesis process. Tumor cells function under lower physiological oxygen levels, known as hypoxia, and they have developed a number of mechanisms to promote angiogenesis and cell survival. Overexpression of HIF-1 in solid tumors is detrimental to tumor suppression and selective inhibition of HIF-1 is an important objective in the search for a chemotherapeutic agent against solid tumors. Dervan and colleagues' polyamides[32] of *N*-methylimidazole and *N*-methylpyrrole amino acids are able to recognize specific DNA sequences by specific pairing with the heterocyclic amino acids, however, these will not be reviewed here as they are not small-molecule inhibitors. The natural product echinomycin,[33] which is a DNA-binding, bisintercalating compound identified as a HIF-1-DNA inhibitor will also not be reviewed here for the same reason. The potent natural product Manassantin A (**58a**) possessing eight chiral centers and a symmetrical tetrasubstituted tetrahydrofyran core was chosen for discussion.

The Manassantin group of related dineolignan natural products was isolated by Rao and Alvarez in 1983 from methanol extracts of *Saururus cernunus* L. (from Greek: saurus = lizard, and oura = tail), commonly known as lizard's tail.[34] This fragrant ornamental and medicinal plant is native to eastern North America, and grows in wet areas and shallow water. Rao and Alvarez isolated and partially characterized Manassantin A (**58a**) and Manassantin B (**58b**) (Figure 1.5), however, their absolute configuration was only recently determined.[35]

Lee and colleagues have also isolated **58a** and **58b** from the roots of *Saururus chinensis,* a perennial medicinal herb found in China and Korea, which has been used in folk medicine to treat edema and gonorrhea.[36] The yield of **58b** isolated from the dried leaves and stems of *S. cernuus* was only 0.1%, hence a total synthesis was required for studying further analogues of these compounds. Manassantins A and B inhibit HIF-1 in the low nanomolar region and, furthermore, inhibit PC-3 prostate cancer cell growth and that of other cell lines with IC_{50} values superior to those for cisplatin and doxorubicin. The molecular weights of these natural products are 732 and 716 respectively, which does not allow their classification as small molecules, however, they are

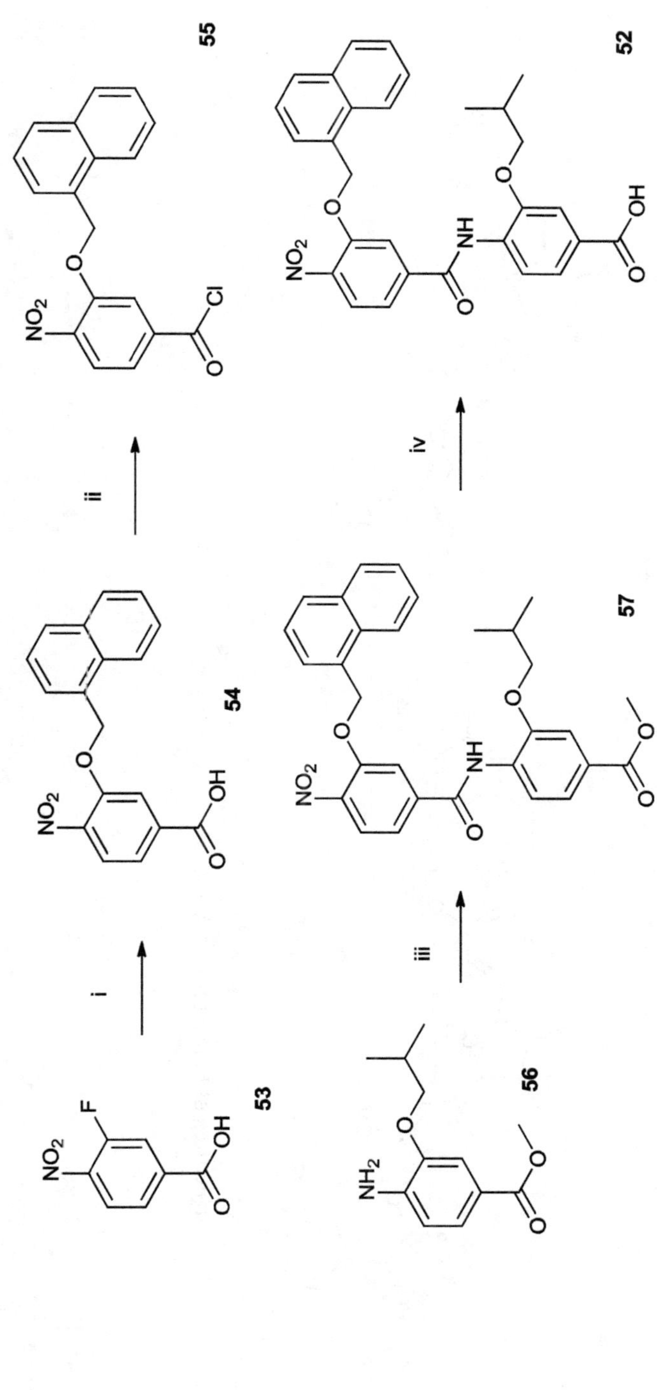

Scheme 1.7 Reagents and conditions: (i) ROH (1.3 equiv.), NaH (3 equiv.), THF; (ii) oxalyl chloride, DMF (catalyst), DCM; (iii) **55**, *N*,*N*-dimethylaniline, acetone; (iv) NaOH, H_2O, MeOH, THF.[31]
Reprinted with permission from K.-Y. Jung, H. Wang, P. Teriete, J. L. Yap, L. Chen, M. E. Lanning, A. Hu, L. J. Lambert, T. Holien, A. Sundan, N. D. P. Cosford, E. V. Prochownik and S. Fletcher, *J. Med. Chem.*, 2015, **58**, 3002. Copyright 2015, American Chemical Society.

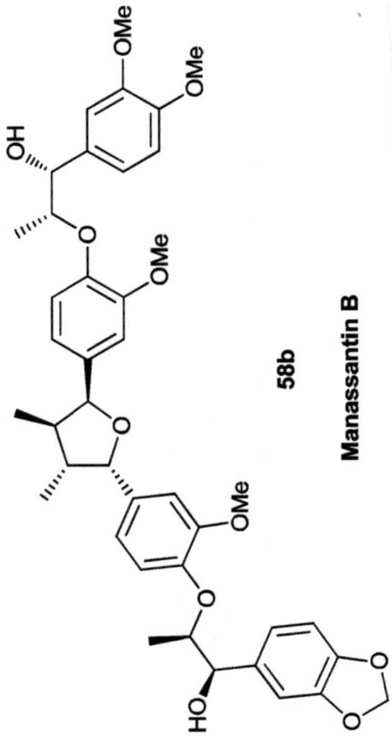

58b

Manassantin B

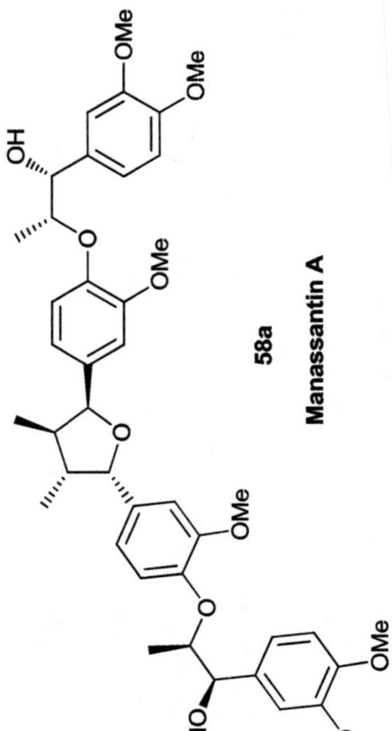

58a

Manassantin A

Figure 1.5 Structures of Manassantins A and B.

clearly not macromolecules. This, together with their interesting biological activities, and low availability makes them interesting targets for total synthesis. There have been two total syntheses of **58a** and **58b**, one by Hanessian and colleagues,[37] and one by Hong and colleagues.[38a,b,39] Furthermore, there has also been a recent patent claiming the synthesis of various synthetic analogues.[38c] The method of synthesis of **58a** used by Hong and colleagues is outlined in Scheme 1.8, and started with the stereoselective Evans asymmetric *syn*-aldol reaction of (*S*)-4-isopropyl-3-propionyloxazolidin-2-one **59** with commercially available 4-benzyloxy-3-methoxybenzaldehyde (88%), followed by protection of the hydroxy group with TBDMS-Cl to give **60** (91%) establishing thus the absolute configuration of two of the asymmetric centers of the natural products. Reductive removal of the chiral auxiliary with sodium borohydride gave the primary alcohol **61** with 88% yield. The latter was converted to the mesylate and then homologated with sodium cyanide to provide nitrile **62** (89% for the two steps). The nitrile was hydrolyzed simultaneously with deprotection of the TBMS ether using sodium hydroxide in refluxing THF, and then the mixture was acidified with HCl in ether to give the γ-lactone **63** at 70% yield.[38b] The next stage in the synthesis, the introduction of the remaining two substituents on the lactone ring of **63** to generate the core tetrasubstituted tetrahydrofuran ring of **58a** and **58b** represented a particular challenge. The challenge was however overcome by the introduction of an *exo*-methylene group instead of the required methyl group, and a subsequent hydrogenation after the introduction of the second substituent.

Lactone **63** was treated with LiHMDS, followed by reaction of the resulting enolate with Eschenmoser's salt, and then oxidation with *m*-CPBA to provide the *exo*-methylene **64** in 80% (for the two steps). The lactone was reduced with DIBAL-H at −78 °C (88%), and the resulting lactol was then treated with $PhSO_2H$ in the presence of $CaCl_2$ to give sulfone **65** (64%). The sulfone was displaced with (4-benzyloxy-3-methoxyphenyl)zinc(II) bromide to give **66** at 41% (dr 10 : 1). Asymmetric hydrogenation of **66** in the presence of Ir and (4*S*,5*S*)-ThrePHOX provided **67** at 99% yield (dr 4 : 1).[40] This established the required geometry around the tetrahydrofuran ring, confirmed by ^1H nuclear magnetic resonance as 2,3-*cis*-3,4-*trans*-4,5-*cis*. The formation of **66** with a dr of 10 : 1 was an improvement over the direct introduction of a methyl group by alkylation of **63**, followed by the same sequence of DIBAL-H reduction, sulfone formation and organozinc displacement where the ratio was only 2 : 1. An additional challenge was the reduction of the *exo*-methylene group of **66**, where catalytic hydrogenation using Pd/C, Adam's catalyst or diimide gave **67** as the minor product. This challenge was overcome by the use of the Ir catalyst, which proceeded with 99% yield and a dr of 4 : 1.

The benzyl ether was hydrogenolyzed over Pd/C to produce the phenolic product **68** (95%) and the latter was reacted on both hydroxy groups with 4 equiv tosylate **69** in the presence of a phosphazene base (BEMP) to give **70** (93%).[37] This S_N2 reaction was far superior to Mitsunobu coupling or oxidation–reduction protocols. The synthesis of Manassantin A (**58a**) was completed with 85% yield following a stereoselective double reduction of the two carbonyl groups with polymer-supported borohydride.

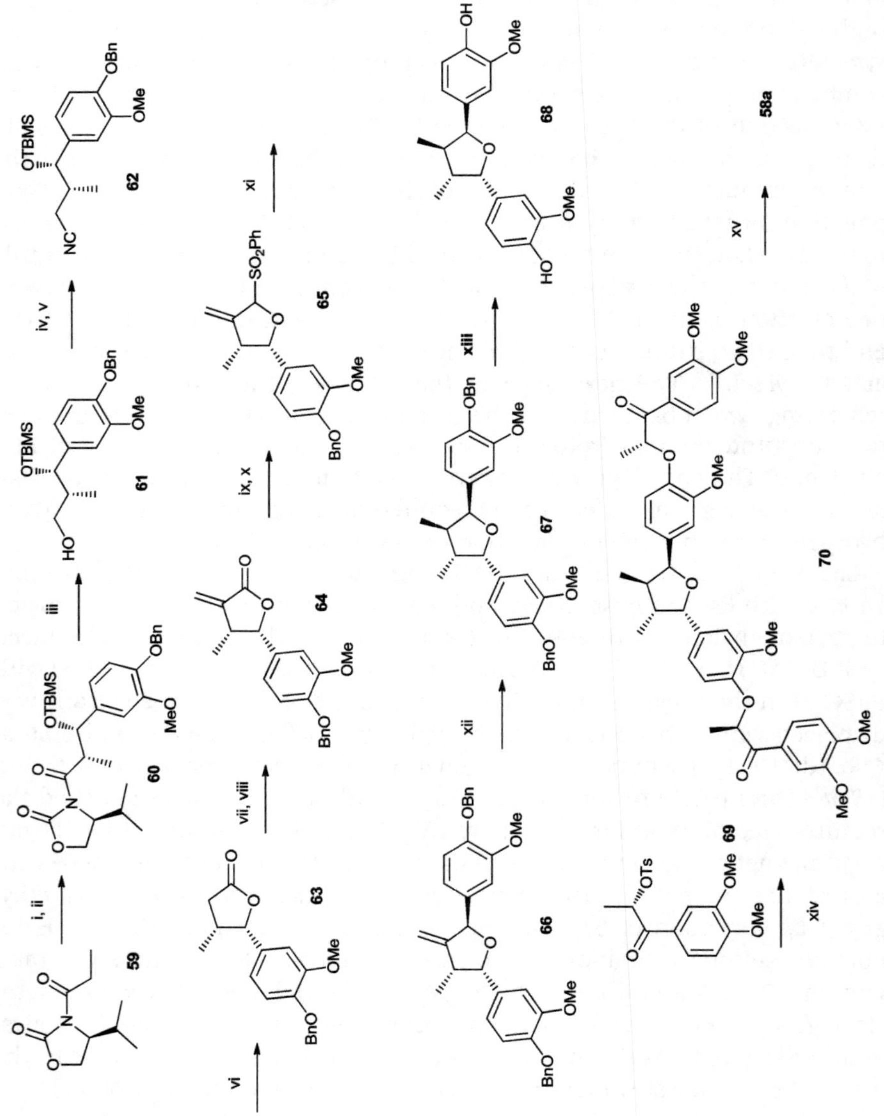

1.6 Inhibitors of NF-κB and NF-Y

The nuclear factor kappa-light chain-enhancer of activated B cells (NF-κB) is a transcription factor involved in cellular responses to stimuli such as cytokines, UV irradiation, stress, free radicals, *etc.* It plays a critical role in regulating the immune response to infection, hence not only has incorrect regulation of NF-κB been associated with cancer, but also autoimmune diseases, inflammation, viral infection and more. The heterotrimeric transcription factor NF-Y is involved in cell differentiation and cell cycle progression; it cooperates with other transcription factors and regulates promoter activity. Once again, the interaction between NF-Y and DNA has been identified as a potential target, wherein blocking of this function could be a strategy for stopping the proliferation of tumours.[41a] Several inhibitors of these transcription factors have been identified, for example resveratol[41b] (this will not be discussed as it is readily available from several plants and, due to its simplicity, its synthesis is of little interest) and hairpin polyamides[1b] (similarly uninteresting synthetically). Some very potent inhibitors of NF-κB and NF-Y that have been discovered belong to a family of compounds known as pyrrolobenzodiazepines (PBDs), whose syntheses form the basis of this section. Since their discovery in the 1960s,[41c] there have been myriad reports of both natural and synthetic PBDs that have shown promise as bioactive compounds of value. A general structure **71** is shown in Figure 1.6, with the typical imine at C11, where a guanine-DNA typically binds covalently within the minor groove of the helix.

The success of PBDs is evidenced, for example, by the NF-κB inhibitor KMR-28-39 (**72a**) (Figure 1.7) (which has impressively exhibited up to femtomolar effectiveness against breast cancer cell lines), GWL-78 (**72b**),[41a] which has been shown to be capable of inhibiting NF-Y, and SJG-136 (**73**) (Figure 1.8), a PBD 'dimer' that has thus far progressed to Phase II clinical trials.[42]

It is worthy of note that PBDs are considered to be interchangeably equivalent in terms of their imine, carbinolamine (hemiaminal) and carbinolamine-ether forms (Scheme 1.9). Depending on their work-up

---

Scheme 1.8 Reagents and conditions: (i) nBu_2BOTf, Et_3N, DCM, −78 to −40 °C, 4-benzyloxy-3-methoxybenzaldehyde, −78 °C, 88%; (ii) TBMS-Cl, imidazole, DMF, 91%; (iii) $NaBH_4$, THF, H_2O, MeOH, 88%; (iv) MsCl, Et_3N, DCM; (v) NaCN, DMSO, 70 °C, 89% (for the two steps); (vi) NaOH, THF, MeOH, H_2O, reflux, HCl, Et_2O, 70%; (vii) LiHMDS, −78 °C, $Me_2N^+{=}CH_2I^-$, −78 °C; (viii) *m*-CPBA, $NaHCO_3$, THF, 80% (for the two steps); (ix) DIBAL-H, DCM, −78 °C, 88%; (x) $PhSO_2H$, $CaCl_2$, DCM, 64%; xi) ArMgBr, $ZnBr_2$, THF, 41% (10:1); (xii) H_2, Ir-(4*S*,5*S*)-ThePHOX, 99%; (xiii) H_2, Pd/C, EtOAc, EtOH, 95%; (xiv) **69**, BEMP (phosphazene base), DCM, 93%; (xv) (polystyrylmethyl)trimethylammonium borohydride, MeOH, 85%.[38]

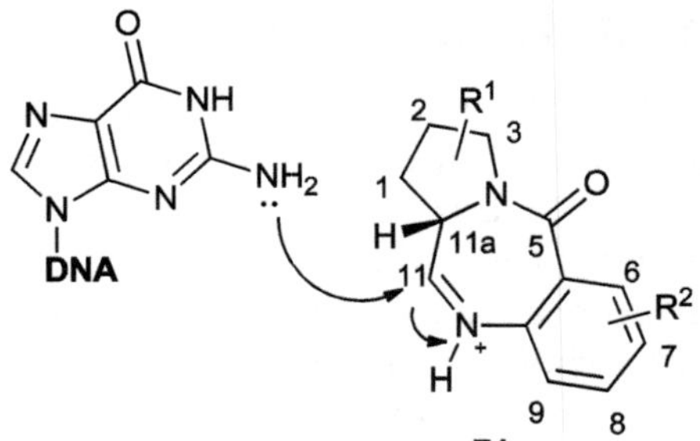

71

Figure 1.6 General structure of PBDs and their mode of binding to guanine-DNA.

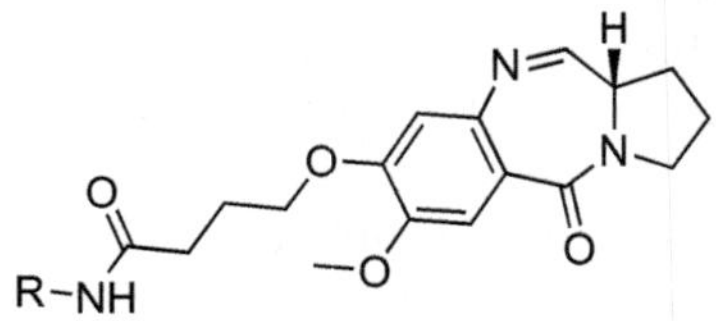

R = **72**

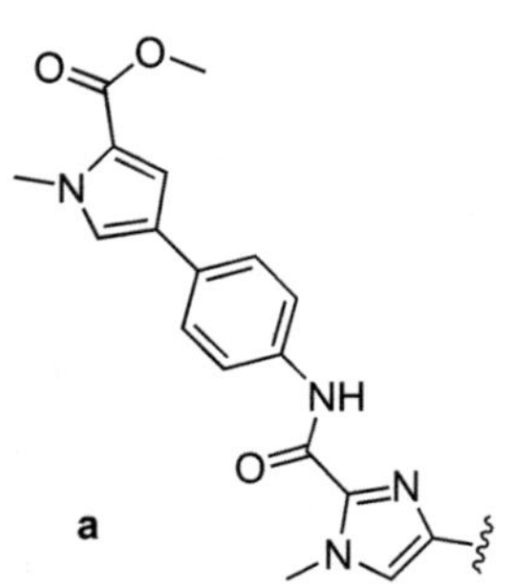

a

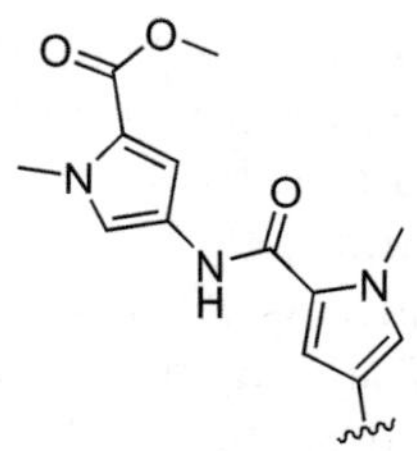

b

Figure 1.7 KMR-28-39 (**72a**), a PBD that inhibits NF-κB, and GWL-78 (**72b**), an NF-Y inhibitor.

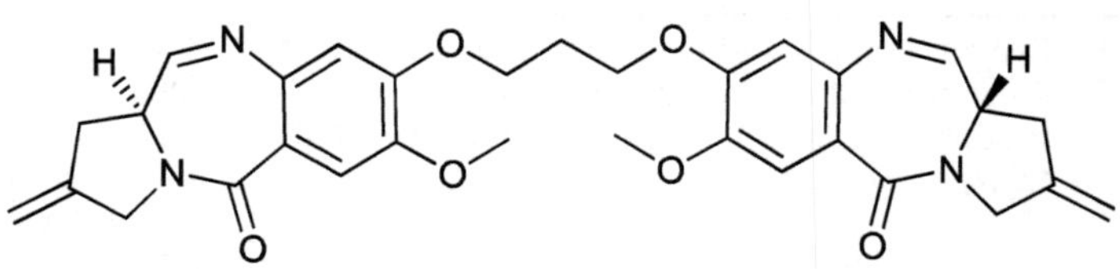

73

Figure 1.8 SJG-136, an example of a dimeric PBD, has progressed to Phase II clinical trials.

procedure, isolation and/or substituents, PBDs can exist as a mixture of forms or exclusively as one.[43] Since testing is typically carried out in an aqueous environment, it is suspected that they will exist as the

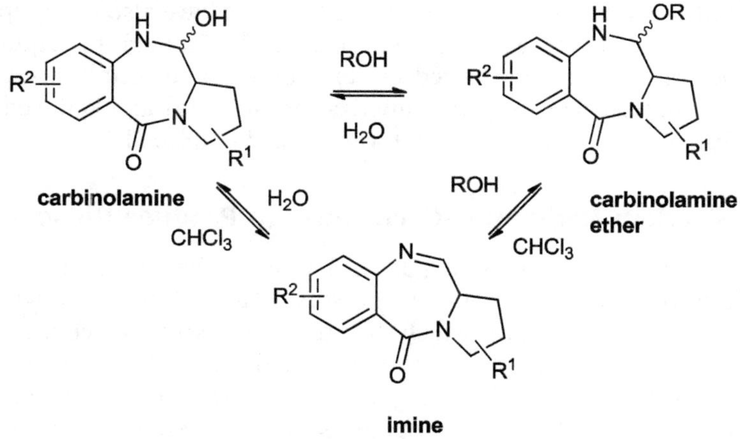

Scheme 1.9 Three equivalent forms of PBDs.

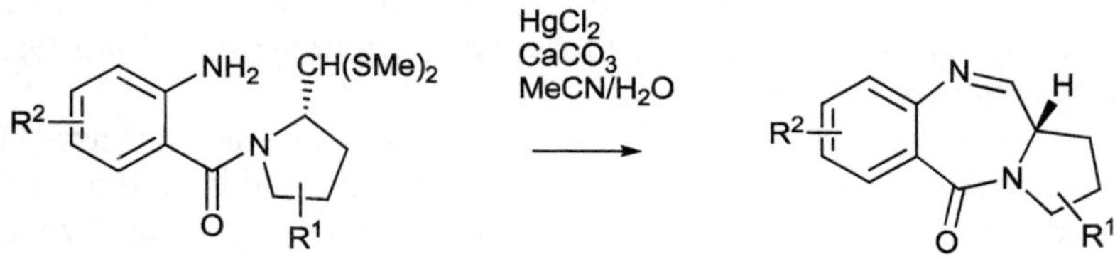

74 **71**

Scheme 1.10 Thioacetal deprotection and *in situ* cyclisation to produce PBDs.

corresponding carbinolamine under physiological conditions, although the mechanism of DNA alkylation has not been fully elucidated.

The C11a stereogenic center (specifically with the (*S*)-configuration) is considered to be crucial to activity as it provides a slight curvature that enables PBDs to slot effectively into the minor groove of DNA, hence L-proline derivatives have proved invaluable precursors in the syntheses of PBDs.

The chemistry of PBDs was comprehensively reviewed by Antonow and Thurston in 2011 and Thurston and Bose in 1994,[43] which covered the various approaches that have been utilized to furnish the characteristic tricyclic core. The main approaches are summarized herein, together with more recent contributions published since 2011.

1.6.1 Tandem Deprotection–Cyclisation Strategy to Produce PBDs

A particularly common approach to PBDs is the mercuric chloride-mediated cyclisation of thioacetal precursors **74** (Scheme 1.10). This is itself obtained from a straightforward amide coupling reaction.

The use of a highly toxic reagent for this key deprotection/cyclisation step however is, of course, less than ideal, and attempts at using the corresponding dimethyl acetal as a safer alternative have also been reported (as well as alternative reagents for deprotection).[43] Although the equivalent acetal species may be deprotected under mildly acidic conditions, a degradation of optical activity (*via* epimerization at C11a) and reduced cytotoxicity was observed when an acetal was instead employed.[43b]

1.6.2 Tandem Reduction–Cyclisation to Produce PBDs

Another approach to synthesizing PBDs has been the reduction of either aromatic nitro- or azide- cyclization precursors (**75** and **76**, respectively) to give the corresponding free amine, which then undergoes an *in situ* condensation to give either the corresponding carbinolamine or imine **71** (Scheme 1.11).

RANEY®-Ni was employed by Langlois *et al.*[44] to reduce the nitro group of analogues **75**, giving PBD-type compounds **71** at moderate to low yield, although this did enable functionalization at C2 that was not previously feasible with the thioacetal deprotection strategy described above.[43a]

Aside from the nitro group reduction approach, azide derivatives **77** have also successfully been cyclized to PBDs using triphenylphosphine (*via* an iminophosphorane intermediate **78**, Scheme 1.12).[45]

A remarkable 98% yield was obtained for this intramolecular aza-Wittig cyclization step, however, azides are not considered to be ideal due to their toxicity and potentially explosive nature. Furthermore, this transformation leads to the formation of triphenylphosphine oxide, ($Ph_3P{=}O$), which is often a troublesome by-product to remove from a reaction mixture, especially at larger scales.

1.6.3 Oxidative Approaches to Produce PBDs

One of the more versatile approaches to producing PBDs has been the oxidation of protected amino-alcohols (type **79**), which concomitantly leads to the ring-closed adduct **80** and is then followed by a deprotection reaction to yield the desired imine species **71** (Scheme 1.13).[46]

Various standard oxidation reagents have proved successful, especially 2,2,6,6-tetramethylpiperidin-1-yl)oxyl (TEMPO) in combination with a hypervalent iodine source, usually phenyliodine diacetate (PIDA), as well as Swern conditions.[46] Common protecting groups (PG) utilized in this procedure include the "Alloc" and "Troc" groups (allyl and trichloroethyl carbamates, respectively). These were deprotected using typical palladium(0) conditions in the case of the Alloc group, and a cadmium–lead couple for the Troc group, and leading to concomitant formation of imine species **71**. Although this approach provided the desired PBDs with typically high yields, the deprotection step tends to be the final step of the synthesis of PBDs, which is not always strategically sound. Moreover, the removal of toxic heavy metals from the final step in the synthetic route of a potential drug is problematic at best.

Another oxidative process involves the synthesis of the corresponding saturated amine PBD-type molecules **81**, followed by treatment with hypervalent iodine reagent, phenyliodine bis(trifluoroacetate) (PIFA) (Scheme 1.14).[47]

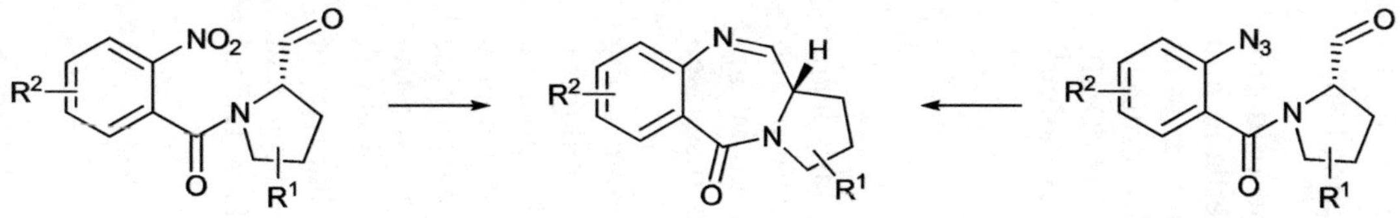

75 71 76

Scheme 1.11 Reduction/cyclisation approaches to produce PBDs.

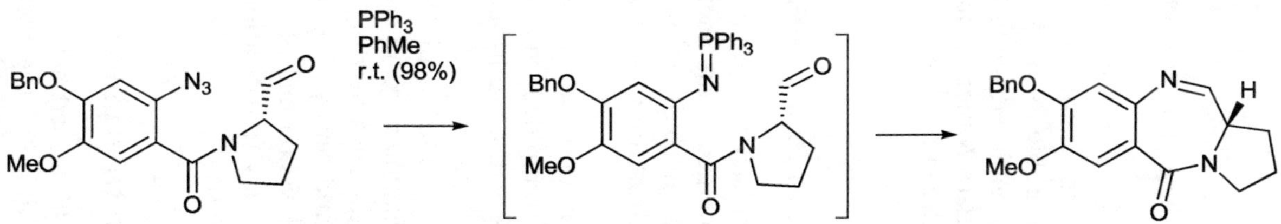

77 78 71

Scheme 1.12 Synthesis of PBD *via* a tandem Staudinger/aza-Wittig reaction.

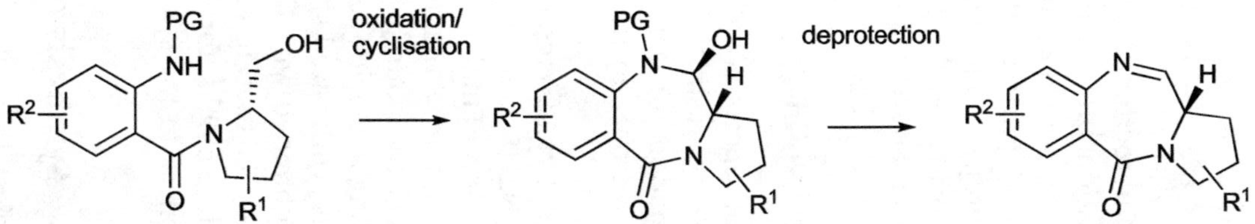

79 80 71

Scheme 1.13 The oxidation–cyclisation followed by deprotection approach.

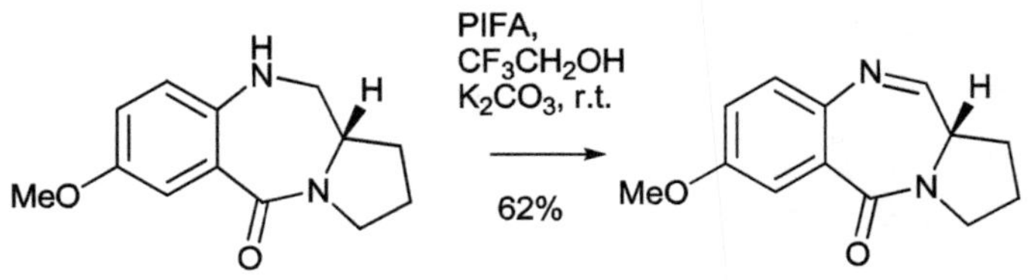

Scheme 1.14 Hypervalent iodine-mediated oxidative process for producing PBDs.

Trifluoroethanol is employed as solvent in this case, as the oxidative cyclisation process did not occur in acetonitrile or in nitromethane and although not expensive, it is undesirable for large-scale reactions due to its toxicity. Similar results have been achieved using the Ley–Griffith protocol (tetrapropylammonium perruthenate–*N*-methylmorpholine-*N*-oxide TPAP–NMO oxidation,[48] a rhodium-catalyzed *tert*-butyl hydroperoxide oxidation[49] and even Swern conditions.[50]

1.6.4 Solid-Phase Synthesis to Produce PBDs

Several reports of solid-phase chemistry relating to PBDs, particularly using the Wang resin,[43] have also emerged. This facilitates purification and automation for parallel synthesis in conjunction with combinatorial approaches but, on the whole, employs the same transformations discussed hitherto.[51]

Kamal *et al.* employed a resin-bound ester **82** which was cleaved using $AlCl_3$–NaI (Scheme 1.15), followed by amide formation and tandem reductive cyclisation. Final compounds were then purified using solid-supported liquid–liquid extraction (SLE) and pleasingly, enabled the synthesis of a series of compounds from a single key intermediate.[52] Resin cleavage, amide coupling and azido reductive-cyclisation were all enacted in one pot and the imines synthesized showed potential DNA-binding affinity *in vitro*. Products were obtained with decent yield and purity. The advantage of this type of strategy is that it enables rapid validation of synthetic sequences and building blocks for library synthesis (purification especially is a typical bottleneck in the synthesis of libraries).

1.6.5 Synthesis of PBD-type Dilactams

It is clear that dilactams of type **85** are attractive PBD-type intermediates, given the known stability of amides and the plethora of reliable amide-coupling methodologies and reagents available for their synthesis. Indeed dilactams **85** have been prepared *via* a multitude of approaches, most not dissimilar to the conditions already mentioned for the syntheses of PBDs (*vide supra*) including the reduction of nitro or azido groups (**84** and **86**, respectively), with simultaneous intramolecular cyclisation of the resulting aniline with the ester group (Scheme 1.16).[53]

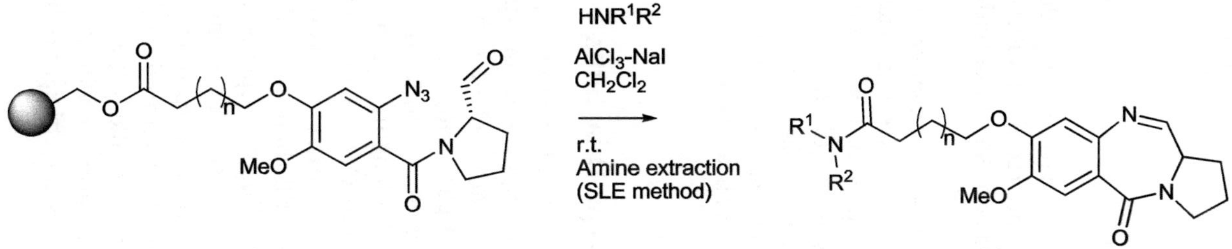

82 83

Scheme 1.15 Example of solid-phase approach for producing PBD libraries.

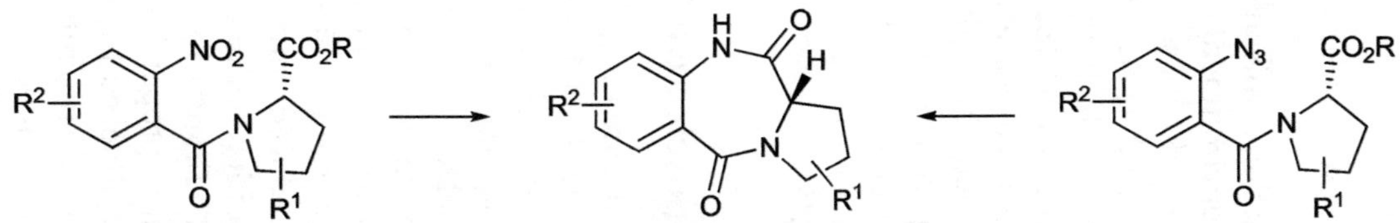

84 85 86

Scheme 1.16 Synthesis of PBD-type dilactams by reduction.

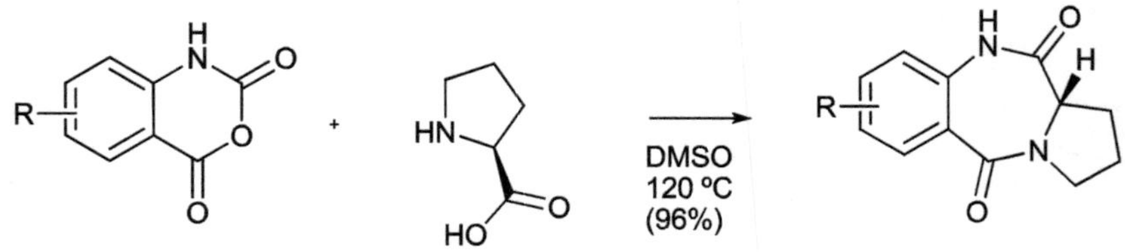

87 **85**

Scheme 1.17 Cyclocondensation to produce dilactams.

Both reductions typically proceed with excellent yield (over 70%), although once again, environmentally undesirable reagents are required (either heavy metals, or in the case of the synthesis of azide **86**, hexamethylphosphoramide, HMPA, an extremely toxic solvent.[43]

Alternatively, dilactams **85** were also made by a cyclocondensation of isatoic anhydrides **87** (Scheme 1.17). This is an attractive approach as it employs inexpensive precursors and yields the expected dilactam with excellent turnover, although elevated temperatures are required.[54]

Lastly, hypervalent iodine reagents have also been employed to synthesize dilactams **90** (Scheme 1.18). This proceeds through a reactive *N*-acylnitrenium intermediate **89** to provide the desired dilactam with a good (up to 70%) overall yield.[55]

1.6.6 Regioselective Reduction of PBD-type Dilactams

Since dilactams are incapable of forming a covalent bond between a guanine of DNA and the N10-C11 imine moiety of PBDs **71** (Figure 1.6), an expedient method that can generate an N10–C11 imine or carbinolamine from the corresponding amide of the dilactam species is of paramount value.

The regioselective reduction of dilactam **85** (Scheme 1.19) to the desired PBD carbinolamine or imine **71** is a particularly challenging endeavor.

Reduction must be preferential towards the N10–C11 amide and must stop at the partially-reduced intermediate carbinolamine/imine as opposed to the fully-reduced corresponding amine. Worse still than over-reduction, harsher conditions have even led to the corresponding ring-opened amino-alcohols.[56] The extent to which these amides are reduced is determined by their electronic properties; evidently, a more electrophilic carbonyl will be more readily attacked by a hydride. Hence, the more electron-donating the aromatic ring substituents or *N*-substituents are, the less likely it will be that total reduction will occur. Furthermore, steric encumbrance will play a role, as a considerably bulky *N*-protecting group will block the efficient delivery of a hydride.[43]

Total syntheses of anthramycin[57] (**93**) by Stille and colleagues and Leimgruber and colleagues incorporated a benzylidene protection (**91**) followed by reduction with sodium borohydride, yielding the carbinolamine **92** with an impressive quantitative yield (Scheme 1.20).

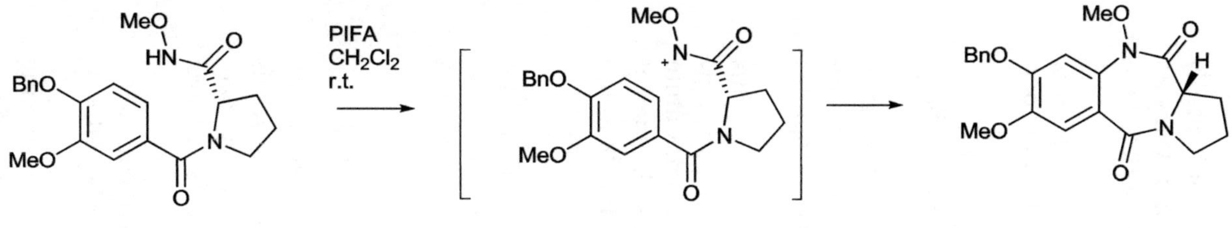

88 **89** **90**

Scheme 1.18 Access to dilactams *via* an *N*-acylnitrenium species.

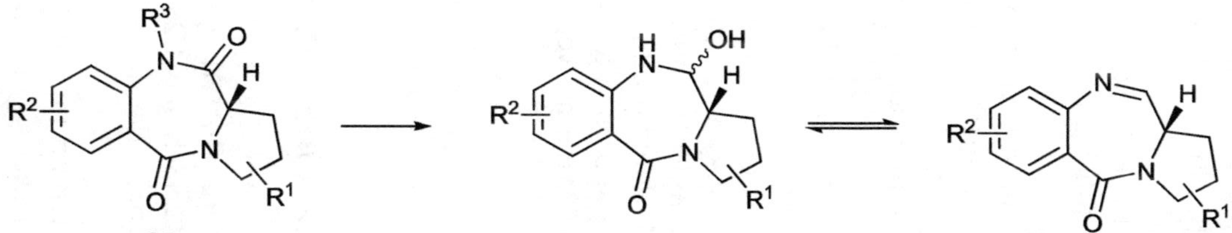

85 **71**

Scheme 1.19 Regioselective reduction of dilactams to produce PBDs.

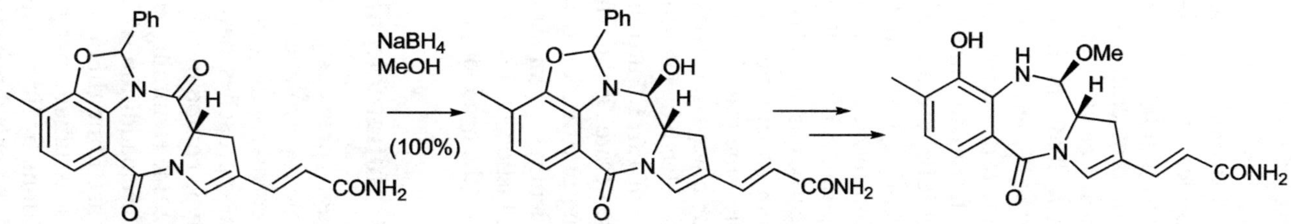

91 **92** **93**

Scheme 1.20 Reduction of a protected dilactam towards the synthesis of anthramycin.

Much more recently, Zemribo and colleagues have adopted the same approach in total syntheses of Limazepine C, D and E (Figure 1.9).[58] Thus, appropriate benzylidene-protected precursors were treated with sodium borohydride, to give the natural products, albeit with disappointingly low yields (13–40%) for this key step.

The benzylidene-protection/reduction approach however is undoubtedly not applicable to all dilactam reductions for the synthesis of PBDs. Many do not feature a hydroxyl group at the C9 position where the appropriate benzylidene protection can be applied.

Howard *et al.*[59] reported a highly efficient reduction of a PBD **94** using initial *N*-SEM amide protection, followed by reduction with lithium borohydride (Scheme 1.21).

Other *N*-protecting groups that have successfully been employed in this type of transformation include the methoxymethyl (MOM) and carboxybenzyl (Cbz) groups, with varying degrees of success.[60]

Recently, Jung and Fletcher reported an application of the Mitsunobu reaction towards these ends (Scheme 1.22).[61] Thus, PBD-related 1,4-benzodiazepine-2,5-dione **96** was subjected to phosphine reagents and diisopropyl azodicarboxylate (DIAD) in the presence of various alcohols.

Although this promises to be a different approach towards tackling this troublesome transformation, the *N*-alkylated product **97** was unsurprisingly almost always favored in all examples examined. Indeed, the maximum chemoselectivity favoring the *O*-alkylated product **98**, which was obtained at a ratio of 1.6 : 1. Furthermore, this approach towards synthesizing **98** would also generate triphenylphosphine oxide as well as hydrazine by-products, making purification troublesome.

Evidently, a significant challenge remains as more general conditions for the reduction of dilactams to their corresponding PBD imines remain elusive (without the need for protecting groups and/or excessively stringent conditions).

1.6.7 Intramolecular Cycloadditions to Access Novel PBDs

Although previously known to give the 1,2,3-triazolo-fused PBDs and equivalent pyrrolobenzothiadiazepine (PBTD) type compounds **99** as well as their corresponding aziridines **102** (Scheme 1.23), complimentary cycloaddition approaches have recently been developed that give access to novel PBD-type compounds.[62]

In 2014, Hemming *et al.* reported an intramolecular azide/alkene 1,3-dipolar cycloaddition which led to various PBD-type triazolines **104** and aziridine azetidine thiones **106** (Scheme 1.24).[63]

It is assumed that aziridine formation occurs under the more forcing conditions by first cycloaddition followed by extrusion of N_2. Further functionalised aziridines were also prepared (for example, with an ester attached to the aziridine methylene). Only moderate yields were ever recorded, even after 7 days of refluxing in chloroform. Furthermore, since it is known that DNA-imine adduct formation is essential for biological activity (Figure 1.6), it

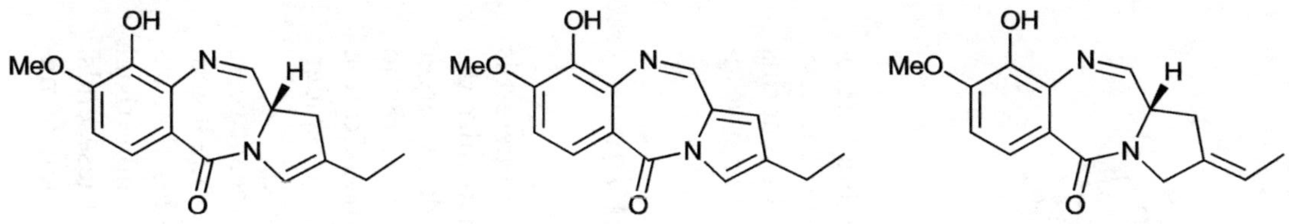

Figure 1.9 Limazepines C, D and E respectively.

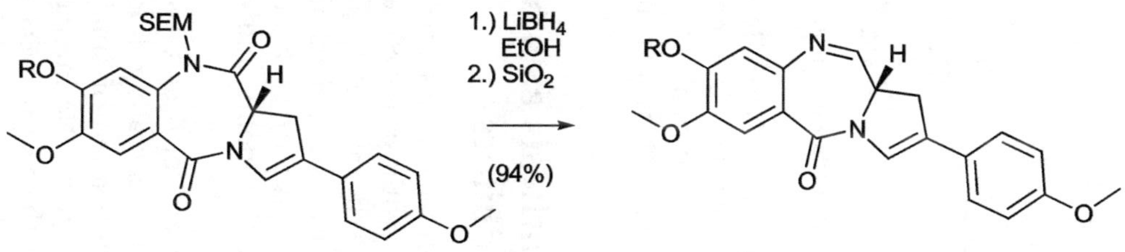

94 **95**

Scheme 1.21 Reduction of *N*-SEM-protected amide to PBD.

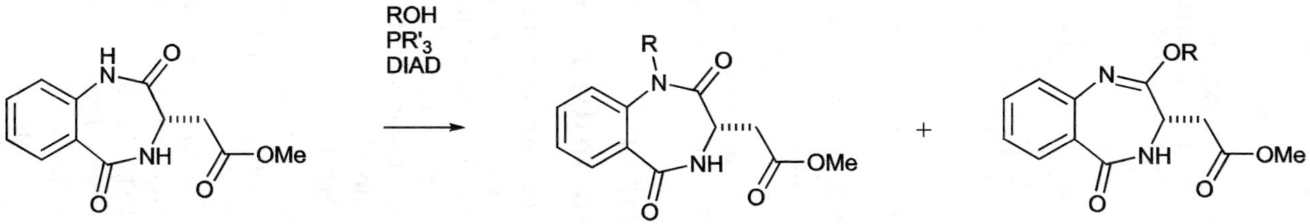

96 **97** **98**

Scheme 1.22 Jung and Fletcher's Mitsunobu reaction towards reduced benzodiazepines.

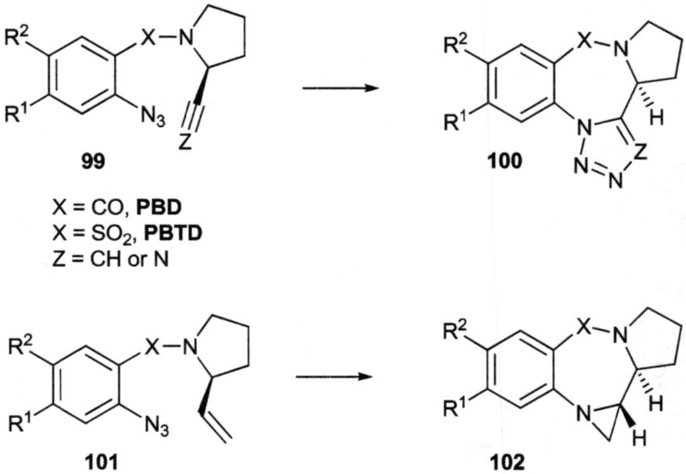

Scheme 1.23 Cycloaddition approaches towards PBDs and PBTDs.

is uncertain whether the replacement of the N10–C11 imine with a heterocycle (compounds of type **104**) will maintain this desired function; aziridines, **106** on the other hand may well be capable of alkylating DNA at this position in a similar way to imine PBDs **71**, which merits further investigation of structures of type **106**.

1.7 Diversity-oriented Synthesis as a Method for Identifying Novel Inhibitors

Contrasted with target-oriented synthesis, where a single target of interest is doggedly pursued, diversity-oriented synthesis (DOS) seeks to maximize the coverage of so-called 'chemical space' by generating an increasingly larger library of compounds that have chemical and skeletal distinctness relative to each other. Using simple transformations only, each structural modification leads to another potential diversification point, and onwards the 'spider-plant' grows (Figure 1.10).[64]

DOS has emerged as a useful concept for drug discovery in recent years and has proved capable of quickly generating libraries for medicinal chemistry programs. Since the identification of small molecules that selectively inhibit protein–protein or protein–DNA interactions is a non-trivial task,[2] several researchers envisaged the use of DOS towards these ends. Indeed DOS has already been employed towards the identification of numerous transcription factor inhibitors.

Among the earliest of such reports (2003), Schreiber and colleagues[65] described a microarray-based screen aimed at finding small molecules that bind Hap3p (heme activator protein homolog 3p), a transcription factor involved in mitochondrial function, which is relevant to several processes, for example in cancer and diabetes. Microarrays containing 12 396 compounds, derived from three different DOS pathways, were prepared in a 'one bead one stock solution' format and then printed onto microscope slides made of chlorinated glass (Scheme 1.25).

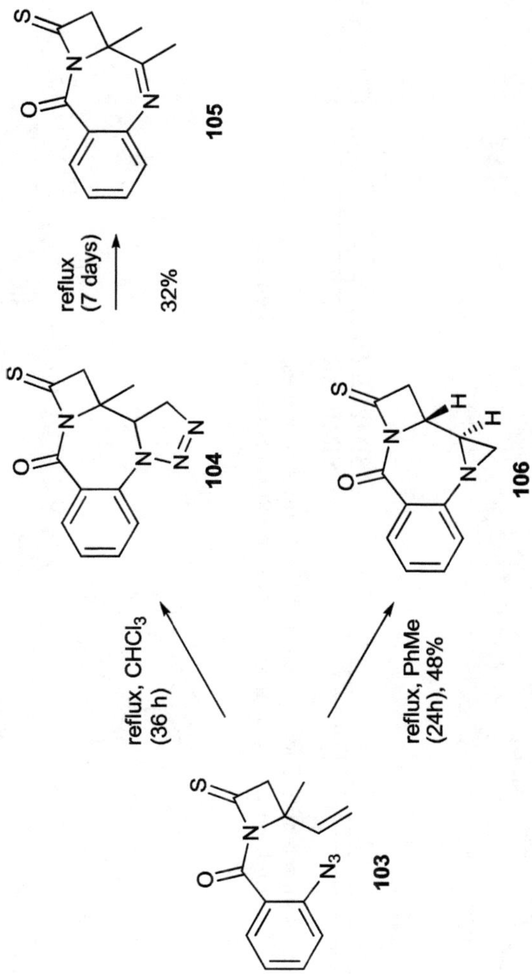

Scheme 1.24 Hemming and colleagues' cycloaddition approach towards producing novel PBD-type frameworks.

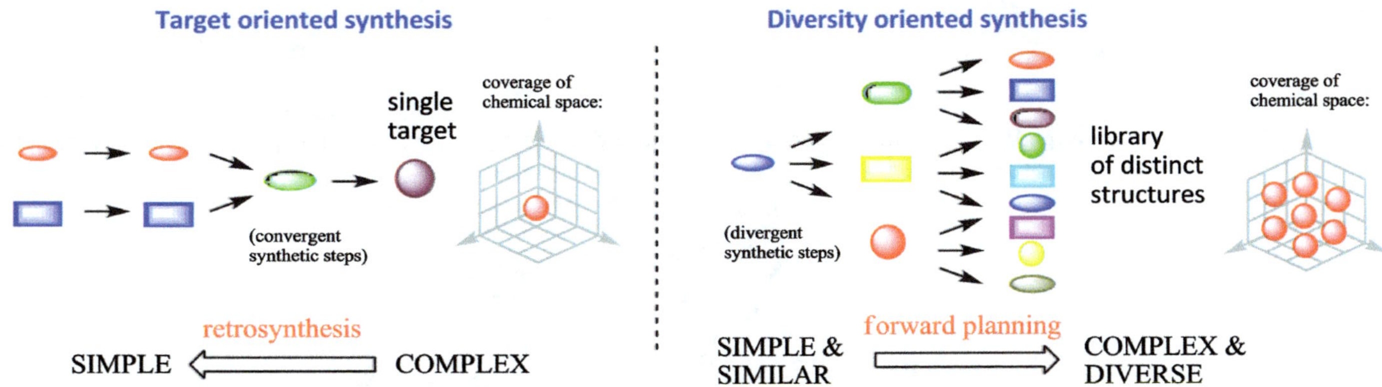

Figure 1.10 Comparison of target- and diversity-oriented synthetic approaches. Reproduced from ref. 64 with permission from The Royal Society of Chemistry.

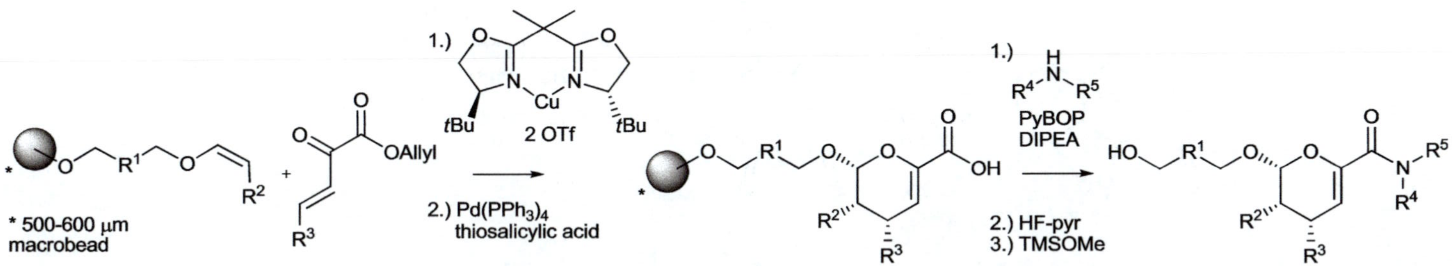

107 **108** **109** **110**

Scheme 1.25 Schreiber and colleagues' DOS approach towards inhibitors of Hap3p.

The three microarrays were then probed with fusion protein Hap3p–glutathione S-transferase (GST) and binding was detected with a cyanine dye Cy5-labelled antibody against the GST section of the protein.

More recently, Shaw and colleagues[66] confirmed the effectiveness of DOS as a starting point when looking for chemical modulators of protein–biopolymer interactions, in this case, those of transcription factors with DNA. Using split–pool techniques, a library of 400 distinct compounds on a solid phase were prepared (Scheme 1.26).

By adopting a *formal* cycloaddition between an imine and an anhydride, stereoselective methods for the preparation of carboxy-γ-lactams (**112** and **113**) and carboxy-2-quinolones (**114** and **115**) were found, hence leading to skeletal, stereochemical and functional group diversity. High diastereoselectivities of up to 95 : 5 were observed. The compounds were then assayed in over 150 high-throughput screens (HTS), and *hit* compounds were readily purified by mass-directed HPLC for further studies. A HTS that employed fluorescence polarization revealed carboxamide **116** (Figure 1.11) to be a novel inhibitor of the interaction of the HOXA13 DNA binding domain to its target sequence of DNA.

HOXA13 is a member of the Hox transcription-factor family, which is required for the normal development of limb and genitourinary tissues. Furthermore, tumor formation is known to closely mimic the processes of development controlled by Hox proteins and these are often abnormally expressed in cancerous tissues. Hence, the successful blocking of Hox protein–DNA interactions could be a viable treatment for the disease. A quick inspection of **116**, however, easily reveals some issues regarding drug-likeness. Lipophilicity associated with the adamantane group, the toxicity associated with the nitro group and the metabolic lability associated with the primary alcohol are areas that require further investigation to optimize lead compounds, such as **116**, found by this valuable hit-generating methodology.

Given the importance of benzodiazepine derivatives in the search for highly potent and selective transcription factor inhibitors (*vide supra*), innovative synthetic approaches for these frameworks are of considerable interest. In 2014, Hajela and colleagues[67] disclosed a new diversity-oriented route towards highly diversified pyrrolo-triazolo benzodiazepines **119** (Scheme 1.27).

Employing a Lewis-acid-promoted C2 selective functionalization and the well known Huisgen azide–alkyne 1,3-dipolar cycloaddition (in one pot), products were produced with excellent yield and using only a very low catalyst loading. This transformation represents the pinnacle of 'step economy' with a significant enhancement in molecular complexity in just one step and was even able to furnish products with a new *spiro* center, regardless of pronounced steric encumbrance in some cases, and always with high yield.

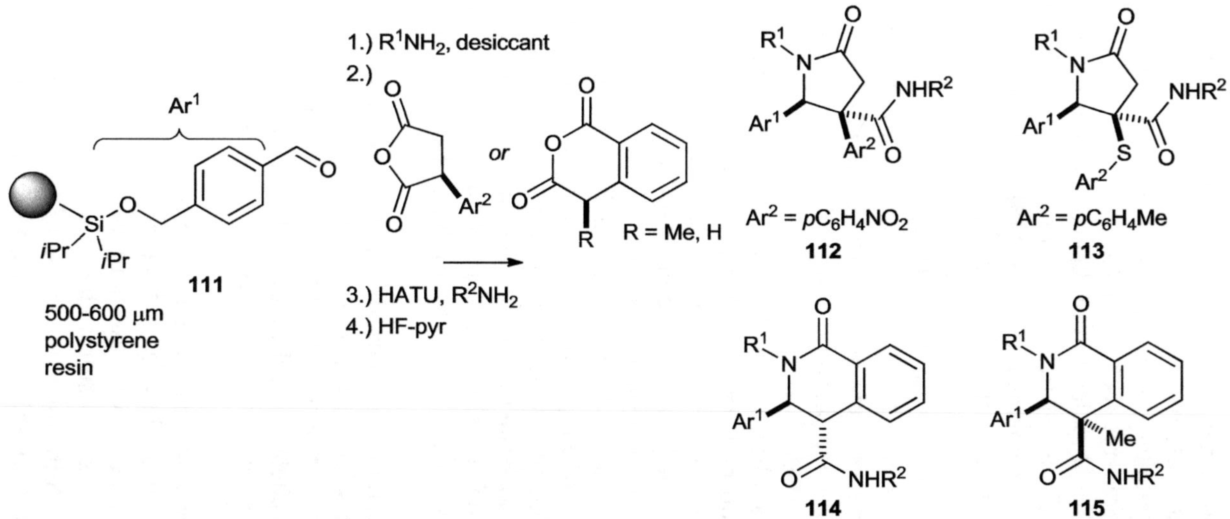

Scheme 1.26 Imine/anhydride formal cycloaddition for library synthesis.

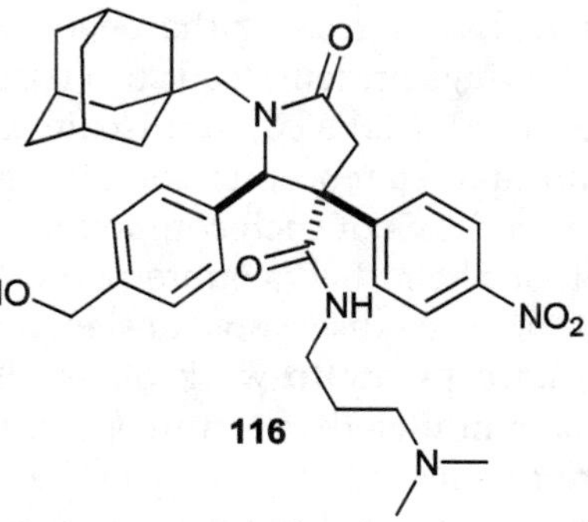

Figure 1.11 Novel HOXA13 transcription factor inhibitor with $IC_{50} = 6.5\ \mu M$.

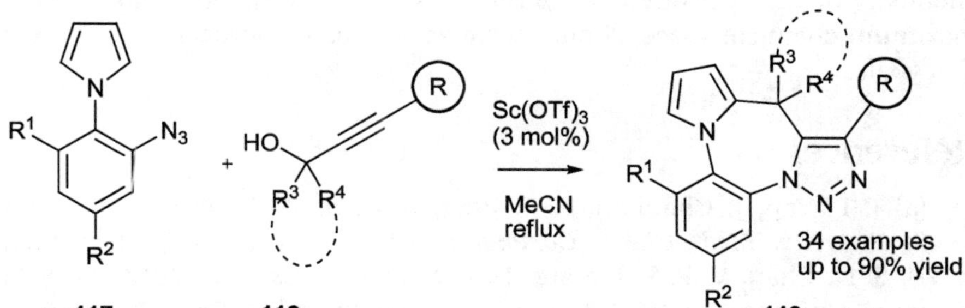

Scheme 1.27 DOS tandem C2-functionalisation/Huisgen-type cycloaddition to produce benzodiazepines.

1.8 Conclusion

Interest in the discovery of various transcription factor inhibitors has increased in recent years, as judged by the large number of publications in this area. Only a few years ago the inhibition of protein–protein interaction by small molecules was considered to be a dream. A variety of inhibitor classes have been identified, ranging from natural products, to pyrrolobenzodiazepines, to peptidomimetics and to diverse libraries that cover a broader range of chemical space. The target proteins, as in the case of c-Myc–Max, do not always have a well defined binding site that can be investigated, hence, the rational design of such inhibitors is more challenging and relies on high-throughput screening of chemical libraries. In all cases though, whether of a natural product, a rationally designed inhibitor or a chemical library, good synthetic chemistry skills for devising a short, convergent and efficient route are paramount for the identification of new medicines. Several modern trends appear to be dominating the ‘zeitgeist’ of organic synthesis, for instance, green chemistry considerations, asymmetric synthesis, C–H activation methodologies and organocatalysis, to name a few. With the well-established medicinal chemistry principles of increased solubility, reduced lipophilicity, increased ligand efficiency and lower predicted dose, coupled to the recent realization that increasing the degree of saturation within a molecule (fewer aromatic rings present) leads to a reduced rate of attrition of candidate drugs in the clinic[68] (reduced promiscuity and toxicity), a renewed interest in sp^3-rich, multiple stereocenter-containing small molecules has emerged.[69] Clearly, the synthesis of such compounds is more expensive and challenging than that of their 'flat' counterparts. This will ensure that chemists at all stages of drug discovery (design, lead optimization, and process development) have plenty to work on in the coming years. This should also serve as a reminder that Nature's compounds are inherently complex and, if we are to succeed in producing better spatial interactions with, for example, target proteins, then we can expect that more complex targets, with a high degree of three-dimensional character, will be needed. Hence, better rationally designed compounds using modern computational chemistry or diversity-oriented synthesis, with emphasis on occupying the maximum chemical space, should serve as adequate approaches to achieve this goal.

References

1. (a) J. L. Yap, J. Chauhan, K.-Y. Jung, L. Chen, E. V. Prochownik and S. Fletcher, *Med. Chem. Commun.*, 2012, **3**, 541; (b) C.-H. Leung, D. S.-H. Chan, V. P.-Y. Ma and D.-L. Ma, *Med. Res. Rev.*, 2013, **33**, 823; (c) M. Gniazdowski, W. A. Denny, S. M. Nelson and M. Czyz, *Expert Opin. Ther. Targets*, 2005, **9**, 471; (d) H.-D. Arndt, *Angew. Chem. Int.*, 2006, **45**, 4552; (e) A. N. Koehler, *Curr. Opin. Chem. Biol.*, 2010, **14**, 331; (f) C. M. Stellrecht and L. S. Chen, *Cancers*, 2011, **3**, 4170; (g) C. Y. Majmudar and A. K. Mapp, *Curr. Opin. Chem. Biol.*, 2005, **9**, 467; (h) T. Berg, *Curr. Opin. Chem. Biol.*, 2008, **12**, 464; (i) M. Gniazdowski, W. A. Denny, S. M. Nelson and M. Czyz, *Curr. Med. Chem.*, 2003, **10**, 909; (j) A. H. Brivanlou and J. E. Darnell Jr., *Science*, 2002, **295**, 813.
2. (a) A. G. Cochran, *Curr. Opin. Chem. Biol.*, 2001, **5**, 654; (b) A. G. Cochran, *Chem. Biol.*, 2000, **7**, R85; (c) J. B. Gibbs, *Science*, 2000, **287**, 1969; (d) A. Abbott, *Nature*, 2002, **416**, 470.
3. (a) A. Nadin, C. Hattotuwagama and I. Churcher, *Angew. Chem., Int. Ed.*, 2012, **51**, 1114; (b) T. Gaich and P. S. Baran, *J. Org. Chem.*, 2010, **75**, 4657.
4. R. G. Naik, S. L. Kattige, S. V. Bhat, B. Alreja, N. J. de Souza and R. H. Rupp, *Tetrahedron*, 1988, **44**, 2081.
5. S. L. Kattige, R. G. Naik, A. D. Lakdawalla, A. N. Dohadwalla, R. H. Rupp and N. J. de Souza, *U. S. Pat.* US4900727, 1990.
6. R. G. Naik, B. Lal, R. H. Rupp, H. H. Sedlacek and G. Dickneite, *U. S. Pat.* US5284856, 1994.
7. K. K. Murthi, M. Dubay, C. McClure, L. Brizuela, M. D. Boisclair, P. J. Worland, M. M. Mansuri and K. Pal, *Bioorg. Med. Chem. Lett.*, 2000, **10**, 1037.
8. K. S. Kim, J. S. Sack, J. S. Tokarski, L. Qian, S. T. Chao, L. Leith, Y. F. Kelly, R. N. Misra, J. T. Hunt, S. D. Kimball, W. G. Humphreys,

B. S. Wautlet, J. G. Mulheron and K. R. Webster, *J. Med. Chem.*, 2000, **43**, 4126.
9. A. Gross, D. R. Borcherding, D. Friedrich and J. S. Sabol, *Tetrahedron Lett.*, 2001, **42**, 1631.
10. J. Turkson, J. S. Kim, S. Zhang, J. Yuan, M. Huang, M. Glenn, E. Haura, S. Sebti, A. D. Hamilton and R. Jove, *Mol. Cancer Ther.*, 2004, **3**, 261.
11. V. M. Shahani, P. Yue, S. Fletcher, S. Sharmeen, M. A. Sukhai, D. P. Luu, X. Zhang, H. Sun, W. Zhao, A. D. Schimmer, J. Turkson and P. T. Gunning, *Bioorg. Med. Chem.*, 2011, **19**, 1823.
12. D. R. Coleman, Z. Ren, P. K. Mandal, A. G. Cameron, G. A. Dyer, S. Muranjan, M. Campbell, X. Chen and J. S. McMurray, *J. Med. Chem.*, 2005, **48**, 6661.
13. P. K. Mandal, Z. Ren, X. Chen, C. Xiong and J. S. McMurray, *J. Med. Chem.*, 2009, **52**, 6126.
14. K. Mandal, D. Limbrick, D. R. Coleman, G. A. Dyer, Z. Ren, J. S. Birtwistle, C. Xiong, X. Chen, J. M. Briggs and J. S. McMurray, *J. Med. Chem.*, 2009, **52**, 2429.
15. P. K. Mandal, W. S. Liao and J. S. McMurray, *Org. Lett.*, 2009, **11**, 3394.
16. P. K. Mandal, F. Gao, Z. Lu, Z. Ren, R. Ramesh, J. S. Birtwistle, K. K. Kaluarachchi, X. Chen, R. C. Bast, Jr., W. S. Liao and J. S. McMurray, *J. Med. Chem.*, 2011, **54**, 3549.
17. P. K. Mandal, Z. Ren, X. Chen, K. Kaluarachchi, W. S.-L. Liao and J. S. McMurray, *Int. J. Pept. Res. Ther.*, 2013, **19**, 3.
18. S. De Lombaert, L. Blanchard, L. B. Stamford, D. M. Sperbeck, M. D. Grim, T. M. Jensen and H. R. Rodriguez, *Tetrahedron Lett.*, 1994, **35**, 7513.
19. N. Elumalai, A. Berg, K. Natarajan, A. Scharow and T. Berg, *Angew. Chem. Int. Ed.*, 2015, **54**, 4758.
20. S. Adhikary and M. Eilers, *Nat. Rev. Mol. Cell Biol.*, 2005, **6**, 635.
21. S. Pelengaris, M. Khan and G. Evan, *Nat. Rev. Cancer*, 2002, **2**, 764.
22. C. E. Nesbitt, J. M. Tersak and E. V. Prochovnik, *Oncogene*, 1999, **18**, 3004.
23. H. J. Chung and D. Levens, *Mol. Cell*, 2005, **20**, 157.
24. T. Berg, S. B. Cohen, J. Desharnais, C. Sonderegger, D. J. Maslyar, J. Goldberg, D. L. Boger and P. K. Vogt, *Proc. Natl. Acad. Sci. U. S. A.*, 2002, **99**, 3830.
25. S. K. Nair and S. K. Burley, *Cell*, 2003, **112**, 193.
26. Y. Xu, J. Shi, N. Yamamoto, J. A. Moss, P. K. Vogt and K. D. Janda, *Bioorg. Med. Chem.*, 2006, **14**, 2660.
27. J. Shi, J. S. Stver, L. R. Whiby, P. K. Vogt and D. L. Boger, *Bioorg. Med. Chem. Lett.*, 2009, **19**, 6038.
28. J. L. Yap, H. Wang, A. Hu, J. Chauhan, K.-Y. Jung, R. B. Gharavi, E. V. Prochownik and S. Fletcher, *Bioorg. Med. Chem. Lett.*, 2013, **23**, 370.
29. J. Chauhan, H. Wang, J. L. Yap, P. E. Sabato, A. Hu, E. V. Prochownik and S. Fletcher, *ChemMedChem*, 2014, **9**, 2274.
30. X. Yin, C. Giap, J. S. Lazo and E. V. Prochownik, *Oncogene*, 2003, **22**, 6151.

31. K.-Y. Jung, H. Wang, P. Teriete, J. L. Yap, L. Chen, M. E. Lanning, A. Hu, L. J. Lambert, T. Holien, A. Sundan, N. D. P. Cosford, E. V. Prochownik and S. Fletcher, *J. Med. Chem.*, 2015, **58**, 3002.
32. R. M. Doss, M. A. Marques, S. Foister, D. M. Chenoweth and P. B. Dervan, *J. Am. Chem. Soc.*, 2006, **128**, 9074.
33. D. Kong, E. J. Park, A. G. Stephen, M. Calvani, J. H. Cardellina, A. Monks, R. J. Fisher, R. H. Shoemaker and G. Melillo, *Cancer Res.*, 2005, **65**, 9047.
34. K. V. Rao and F. M. Alvarez, *Tetrahedron Lett.*, 1983, **24**, 4947.
35. T. W. Hodges, C. F. Hossain, Y.-P. Kim, Y.-D. Zhou and D. G. Nagle, *J. Nat. Prod.*, 2004, **67**, 767.
36. B. Y. Hwang, J.-H. Lee, J. B. Nam, Y.-S. Hong and J. J. Lee, *Phytochemistry*, 2003, **64**, 765.
37. S. Hanessian, G. J. Reddy and N. Chahal, *Org. Lett.*, 2006, **8**, 5477.
38. (a) H. Kim, A. C. Kasper, E. J. Moon, Y. Park, C. M. Wooten, M. W. Dewhirst and J. Hong, *Org. Lett.*, 2009, **11**, 89; (b) H. Kim, C. M. Wooten, Y. Park and J. Hong, *Org. Lett.*, 2007, **9**, 3965; (c) J. Hong, M. W. Dewhirst, H. Kim, A. C. Kasper, E. J. Moon, Y. Park and C. M. Wooten, Pat. WO2010059858A1, 2010.
39. A. C. Kasper, E. J. Moon, X. Hu, Y. Park, C. M. Wooten, H. Kim, W. Yang, M. W. Dewhirst and J. Hong, *Bioorg. Med. Chem. Lett.*, 2009, **19**, 3783.
40. S. McIntyre, E. Hoermann, F. Menges, P. Smidt and A. Pfaltz, *Adv. Synth. Catal.*, 2005, **347**, 282.
41. (a) M. Kotecha, J. Kluza, G. Wells, C. C. O'Hare, C. Forni, R. Mantovani, P. W. Howard, P. Morris, D. E. Thurston, J. A. Hartley and D. Hochhauser, *Mol. Cancer Ther.*, 2008, **7**, 1319; (b) Z. Ren, L. Wang, J. Cui, Z. Huoc, J. Xue, H. Cui, Q. Mao and R. Yang, *Pharmazie*, 2013, **68**, 689; (c) M. D. Tendler and S. Korman, *Nature*, 1963, **199**, 501.
42. (a) K. M. Rahman, P. J. M. Jackson, C. H. James, B. P. Basu, J. A. Hartley, M. D. L. Fuente, A. Schatzlein, M. Robson, R. B. Pedley, C. Pepper, K. R. Fox, P. W. Howard and D. E. Thurston, *J. Med. Chem.*, 2013, **56**, 2911; (b) P. J. M. Jackson, C. H. James, T. C. Jenkins, K. M. Rahman and D. E. Thurston, *ACS Chem. Biol.*, 2014, **9**, 2432; (c) J. A. Hartley, A. Hamaguchi, M. Suggitt, S. J. Gregson, D. E. Thurston and P. W. Howard, *Invest. New Drugs*, 2012, **30**, 950; (d) J. Mantaj, D. E. Thurston and K. M. Rahman, *Cancer Res.*, 2013, **73**, 2230.
43. (a) D. Antonow and D. E. Thurston, *Chem. Rev.*, 2011, **111**, 2815; (b) D. E. Thurston and D. S. Bose, *Chem. Rev.*, 1994, **94**, 433.
44. N. Langlois, A. Rojas-Rousseau, C. Gaspard, G. H. Werner, F. Darro and R. Kiss, *J. Med. Chem.*, 2001, **44**, 3754.
45. S. Eguchi, K. Yamashita, Y. Matsushita and A. Kakehi, *J. Org. Chem.*, 1995, **60**, 4006.
46. (a) T. Fukuyama, G. Liu, S. D. Linton, S. C. Lin and H. Nishino, *Tetrahedron Lett.*, 1993, **34**, 2577; (b) S. J. Gregson, P. W. Howard, J. A. Hartley, N. A. Brooks, L. J. Adams, T. C. Jenkins, L. R. Kelland and D. E. Thurston, *J. Med. Chem.*, 2001, **44**, 737.
47. G. A. Kraus and A. Melekhov, *Tetrahedron*, 1998, **54**, 11749.

48. A. Kamal, P. W. Howard, B. S. N. Reddy, B. S. P. Reddy and D. E. Thurston, *Tetrahedron*, 1997, **53**, 3223.
49. H. Choia and M. P. Doyle, *Chem. Commun.*, 2007, **7**, 745.
50. A. Kamal and N. V. Rao, *Chem. Commun.*, 1996, 385.
51. A. Kamal, K. L. Reddy, V. Devaiah, N. Shankaraiah and D. R. Reddy, *Mini-Rev. Med. Chem.*, 2006, **6**, 53.
52. A. Kamal, S. Prabhakar, N. Shankaraiah, N. Markandeya, P. V. Reddy, V. Srinivasulu and M. Sathish, *Tetrahedron Lett.*, 2013, **54**, 4435.
53. (a) H. Tang, G. Zhao, Z. Zhou, Q. Zhou and C. Tang, *Tetrahedron Lett.*, 2006, **47**, 5717; (b) A. Kamal, B. S. P. Reddy and B. S. N. Reddy, *Tetrahedron Lett.*, 1996, **37**, 6803.
54. W.-P. Hu, J.-J. Wang, F.-L. Lin, Y.-C. Lin, S.-R. Lin and M.-H. Hsu, *J. Org. Chem.*, 2001, **66**, 2881.
55. A. Correa, I. Tellitu, E. Dominguez, I. Morreno and R. Sanmartin, *J. Org. Chem.*, 2005, **70**, 2256.
56. D. E. Thurston, P. T. P. Kaumaya and L. H. Hurley, *Tetrahedron Lett.*, 1984, **25**, 2649.
57. (a) M. R. Pena and J. K. Stille, *J. Am. Chem. Soc.*, 1989, **111**, 5417; (b) W. Leimgruber, A. D. Batcho and R. C. Czajkowski, *J. Am. Chem. Soc.*, 1968, **90**, 5641.
58. (a) G. Smits and R. Zemribo, *Org. Lett.*, 2013, **15**, 4406; (b) G. Sakaine, G. Smits and R. Zemribo, *Tetrahedron Lett.*, 2015, **56**, 4767.
59. P. W. Howard, Z. Chen, S. J. Gregson, L. A. Masterson, A. C. Tiberghien, N. Cooper, M. Fang, M. J. Coffils, S. Klee, J. A. Hartley and D. E. Thurston, *Bioorg. Med. Chem. Lett.*, 2009, **19**, 6463.
60. (a) M. Mori, Y. Uozumi, M. Kimura and Y. Ban, *Tetrahedron*, 1986, **42**, 3793; (b) T. Nagasaka and Y. Koseki, *J. Org. Chem.*, 1998, **63**, 6797.
61. K.-Y. Jung and S. Fletcher, *Med. Chem. Commun.*, 2012, **3**, 1160.
62. C. S. Chambers, N. Patel and K. Hemming, *Tetrahedron Lett.*, 2010, **51**, 4859.
63. K. Hemming, C. S. Chambers, F. Jamshaid and P. A. O'Gorman, *Molecules*, 2014, **19**, 16737.
64. (a) D. S. Tan, *Nat. Chem. Biol.*, 2005, **1**, 74; (b) D. R. Spring, *Org. Biomol. Chem.*, 2003, **1**, 3867; (c) M. D. Burke and S. L. Schreiber, *Angew. Chem., Int. Ed.*, 2004, **43**, 46; (d) R. J. Spandl, A. Benderb and D. R. Spring, *Org. Biomol. Chem.*, 2008, **6**, 1149.
65. A. N. Koehler, A. F. Shamji and S. L. Schreiber, *J. Am. Chem. Soc.*, 2003, **125**, 8420.
66. P. Y. Ng, Y. Tang, W. M. Knosp, H. S. Stadler and J. T. Shaw, *Angew. Chem., Int. Ed.*, 2007, **46**, 5352.
67. M. K. Hussain, M. I. Ansari, R. Kant and K. Hajela, *Org. Lett.*, 2014, **16**, 560.
68. (a) F. Lovering, J. Bikker and C. Humblet, *J. Med. Chem.*, 2009, **52**, 6752; (b) F. Lovering, *Med. Chem. Commun.*, 2013, **4**, 515.
69. S. L. Schreiber, *Science*, 2000, **287**, 1964.

CHAPTER 2

Computational Approaches in the Development of Small-molecule Transcription Factor Inhibitors

PAUL J. M. JACKSON,*[a] SHIRIN JAMSHIDI*[b] AND DOAA B. FARAG[b,c]

[a] Transcriptogen Ltd, London, UK, www.transcriptogen.com; [b] Institute of Pharmaceutical Sciences, Faculty of Life Sciences and Medicine, King's College, London, UK; [c] Faculty of Pharmacy, Misr International University, Cairo, Egypt
*Email: paul.jackson@femtogenix.com; shirin.jamshidi@kcl.ac.uk

2.1 Introduction

Both proto-oncogenes and tumour suppressor genes regulate cell division, with proto-oncogenes promoting cell growth and suppressor genes reducing cell growth. There are a variety of signal transduction pathways and a number of stress-response pathways conserved in vertebrates, with transcription factors [for example signal transducer and activator of transcription (STAT), myelocytomatosis viral oncogene (MYC), early growth response 1 (EGR1)] participating at the end of these pathways through the ability to control the upregulation or downregulation of specific genes. Both a TATA-binding protein and basal transcription factors (for example TFIIB and the TFIID complex) bind to a TATA box of a promoter region of DNA, promoting

Drug Discovery Series No. 65
Small-molecule Transcription Factor Inhibitors in Oncology
Edited by Khondaker Miraz Rahman and David E. Thurston

Published by the Royal Society of Chemistry, www.rsc.org

the formation of a pre-initiation complex which, in turn, recruits RNA polymerase to begin the process of transcription. This is a slow process, and can be accelerated through activators, in the form of transcription factors. Activators bind to an enhancer region within the DNA, which can be located thousands of base-pairs away from the original pre-initiation complex. The DNA–activator complex then accelerates transcription through the linkage of the pre-initiation complex and activator complex, resulting in the accelerated recruitment of RNA polymerase. In most cases, a transcription factor is activated through ligand binding, phosphorylation (*e.g.*, in the case of the STAT family) and/or dimerization, and these phases are targeted during drug development.

Molecular modelling techniques allow for the atomic-level evaluation of shape-fit and binding affinity of ligands with a receptor, and can be a useful tool in establishing structure–activity relationships of libraries of ligands, thereby driving the drug discovery process. The field of *in silico* screening has progressed rapidly through technological advancement over the past number of years. The speed of computers has doubled every two years ('Moore's Law'[1]), and the use of specialised graphics processing units (GPUs) and super-computers has allowed the docking of large libraries of molecules and simulation of large macromolecules. This has culminated in the 1 ms all-atom molecular dynamics simulations of bovine pancreatic trypsin inhibitor (BPTI)[2] using a specially built, massively parallel super computer capable of simulating in orders of magnitude greater than standard computer systems.[2] This has resulted in super computers being incorporated into *in silico* procedures and methods, and has provided scientists with the capability of quickly screening large libraries of compounds and therefore identifying potential hits with favourable characteristics (*e.g.*, number of hydrogen bond donors or acceptors or molecule shape). When coupled with detailed structures available through NMR or crystallography, these advanced molecular screening techniques have been adopted in the development of inhibitors of various therapeutically-relevant proteins, particularly those once considered 'undruggable', such as transcription factors.

There are a number of examples of the use of *in silico* techniques in the development of inhibitors of transcription factors. For example, the transcription factor nuclear factor (NF)-κB is involved in in a number of processes including cell cycle regulation, apoptosis, specific gene expression and regulation of cell differentiation.[1] NF-κB is upregulated in a number of cancer cell types (*e.g.*, pancreatic cancer) through chronic stimulation of the Inhibitor of NF-κB kinase (IKK) pathway, and continuous NF-κB activity prevents cancer cells from entering apoptosis. As such, NF-κB is a viable target for anti-cancer drug discovery, and computational methods, such as virtual screening, docking, quantitative structure–activity relationship (QSAR), fragment-based screening and molecular dynamics simulations have all been used to good effect in developing inhibitors of the transcription factor.

A similar approach has been adopted for the development of STAT3 inhibitors. STAT proteins are activated by the phosphorylation of a critical tyrosine residue, which is mediated by growth factor receptor tyrosine kinases, Janus kinases or the Src (sarcoma) family kinases. Upon tyrosine phosphorylation, two STAT monomers dimerise and translocate to the nucleus. They then bind to specific DNA response elements in the promoters of target genes, and induce gene expression. Aberrant activity of one of the family members, STAT3, contributes to both carcinogenesis and tumour progression by upregulating gene expression and promoting dysregulated growth, survival and angiogenesis and modulating immune responses.[3]

As a critical step in STAT activation,[4] the dimerization between two STAT monomers presents an attractive target for preventing STAT3 DNA-binding and transcriptional activity and inhibiting STAT3 biological functions.[3e,5] STAT3 dimerization relies on the reciprocal binding of the SH2 domain of one monomer to the Pro–pTyr–Leu–Lys–Thr–Lys sequence of the other STAT3 monomer, and virtual screening and docking have been used to develop STAT3 inhibitors using this approach.

There are several approaches to STAT3-targeted drug design: (i) inhibitors can be designed to directly target the STAT3-SH2 domain *via* a peptide sequence, peptidomimetics or non-peptidic small molecules; (ii) inhibitors can directly target the DNA-binding domain of STAT3; and (iii) antisense approaches, using oligonucleotides, can be adopted to inhibit STAT3 transcription.[6] The design of inhibitors which directly target the STAT3-SH2 domain is the most commonly used approach using a structure-based, computer-aided drug design strategy.[7]

2.2 NF-κB

Atomic-level analyses of NMR and X-ray crystallography structures of NF-κB indicate that there are two distinct druggable areas of the NF-κB protein; the protein–protein interaction (PPI) and the DNA-binding region (DBR). The NF-κB PPI interface exhibits hydrophobic interaction and hydrogen bonding potential. Two unique hydrogen bonds are present in the NF-κB p50–p65 heterodimer that distinguish it from the p50 and p65 homodimers. The homologous residues, Asp-254 (from p50) and Asn-200 (from p65), form a hydrogen bond between their O-δ and N-δ atoms, which contribute to the stability of the heterodimer. Similarly, a hydrogen bond is formed when N-ε2 from His304 contacts S-γ in Cys-197. In p50, Arg-252, Glu-265, Tyr-267, Cys-270, Arg-305 and Phe-307 also interact with the p65 subunit through hydrogen bonds. These residues contact Asp-243, Glu-211, Arg-198, Val-199, Asn-200, His-245, Val-248, Leu-215, Cys-216 and Asp-217 in p65.[8] These specific amino acids encompass the PPI and can be targeted using rational drug design.

The specific amino acids that are referred to as the DNA binding region of NF-κB (Figure 2.1) are residues Arg-59, Tyr-60, Val-61, Cys-62, Glu-63, Gly-64, Pro-65, Ser-66, His-67, Gly-68, Gly-69, Leu-70 and Pro-71 of subunit p50.[8b,9]

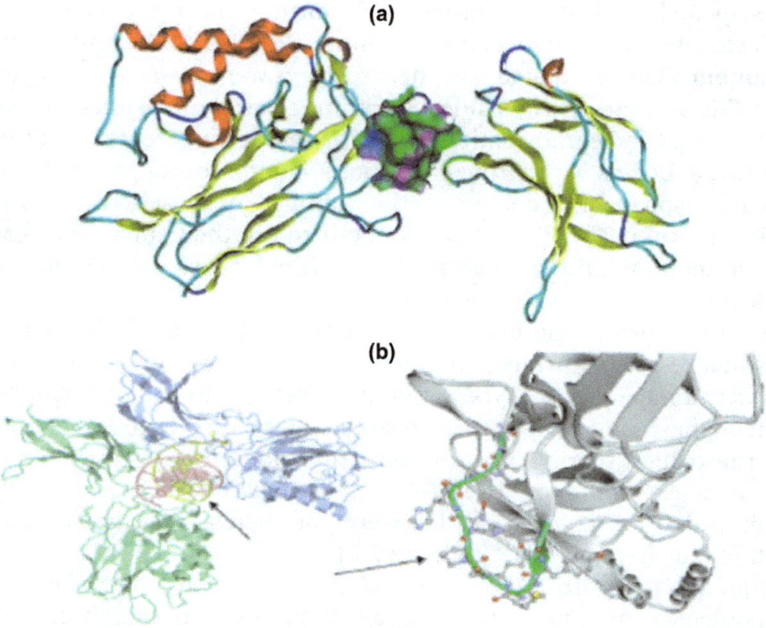

Figure 2.1 (a) Visualization of subunit p50 NF-κB and (b) the DNA binding region of NF-κB, indicated by black arrows (PDB ID 1NFK), green indicates the hydrophobic region of the DBR, while magenta and blue represent hydrophilic and polar regions of the protein.

This region of the p50 subunit has a mainly positive potential and is the first accessible surface for any ligand or the target DNA. As a result, most ligands targeting the DBR are required to have a negative potential, and the calculated molecular electrostatic potential (derived through quantum mechanics calculations)[10] is a valuable resource when developing DBR inhibitor leads.

2.2.1 Docking and Virtual Screening in the Discovery of NF-κB Inhibitors

A number of virtual screening studies have been reported on the NF-κB transcription factor.[11] These studies have all adopted a similar approach, whereby key amino acid residues in either the PPI or DBR were identified and isolated and a large library of molecules were screened against the target using docking software. This was then followed by synthesis or purchase of key hits, and cytotoxicity or transcription factor inhibition analysis. There are a variety of software suites which can be used to support this type of study including DOCK 6,[12] AutoDock[13] and Schrodinger software suites (*e.g.*, Glide[14]).

For example, in one study, the p50 NF-κB subunit was screened and a total of 1484 structures from the ZINC 3D database[15] and six known NF-κB inhibitors [Concentration giving 50% of maximum inhibition (IC_{50}) ≤500 μM] were screened against the protein.[11d] Starting from the co-crystallized complexes, the NF-κB p50–p50 homodimer (PDB ID 1NFK and 1LE9), the p50 monomers were isolated and the structures were prepared for screening using Glide software.[14] The Glide algorithm is based on a systematic search of orientations and conformations of the ligand in the receptor binding site. The software begins with a rough positioning and scoring phase that limits the search space and decreases the number of poses to be selected for minimization. The filtered poses obtained are then subjected to Monte Carlo simulations and the accepted minimized poses are rescored using GlideScore.

The coordinates of the binding site were selected from a set of putative active site residues involved in hydrogen bonding with NF-κB sites of DNA (*i.e.*: Arg-54, Arg-56, Tyr-57, Cys-59, Lys-241, Thr-143 and Gln-306)[11d] and the highest scoring ligands were selected for a further round of docking. 32 of these high-scoring ligands were then identified as hits, and only ZINC database compounds 1797601 (**1**), 1688817 (**2**), 1345112 [(*R*)-**3**], 1345109 [(*S*)-**3**] and 1983981 [(*R*)-**4**] were commercially available and thus chosen for further analysis[11d] (Figure 2.2).

Compound **1** (Figure 2.2) showed the highest G-score (-8.22 kcal mol^{-1}) at physiological pH and outranking all of the docked compounds in the prepared library. Furthermore, electrophoretic mobility shift assay (EMSA) studies revealed that compound **1** inhibited NF-κB to a greater extent than known NF-κB inhibitors, with an IC_{50} of 30 μM. The residues involved in the ligand–protein interactions were from the known DBR and were Lys-144, Lys-145, Thr-143, Tyr-57 and Cys-59 of p50 (chain A) and Lys-145, Lys-146

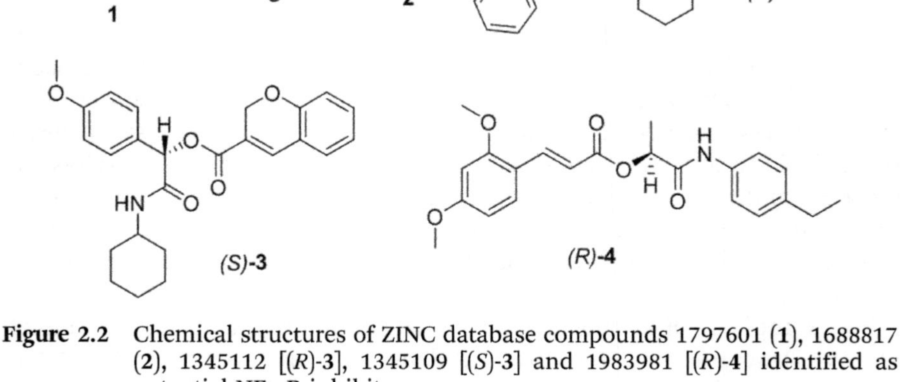

Figure 2.2 Chemical structures of ZINC database compounds 1797601 (**1**), 1688817 (**2**), 1345112 [(*R*)-3], 1345109 [(*S*)-3] and 1983981 [(*R*)-4] identified as potential NF-κB inhibitors.

and Thr-143 of p50 (chain B).[11d] Although moderately effective, the computational studies undertaken allowed the identification of a lead molecule for further derivatisation.

2.2.2 QSAR

QSAR (quantitative structure–activity relationship) is a computational approach whereby chemical structure is quantitatively correlated with either biological activity or chemical reactivity. It allows the determination of core features of libraries of active molecules, and these features are then used as a template to predict activities and features of newly designed molecules. QSAR methods have been used to develop pharmacophore models from literature through the identification of common features of NF-κB inhibitors.[16]

Using this approach, Siedle *et al.* investigated a library of 103 sesquiterpene lactones with six different core scaffolds (all suggested to inhibit NF-κB through alkylation of Cys38 in the DBR) and examined them for common features thought to contribute to their DBR-binding effect.[16a]

The results of QSAR studies indicated that in the case of ligands possessing a rigid skeleton, topological and structure-coding parameters contributed to the NF-κB inhibitory activity of the molecules, whereas inhibition occurred as a result of features such as the number and type of α,β-unsaturated carbonyl structural elements in the case of flexible-skeleton ligands. Further structural features, such as an acyl moiety located near to an exocylic methylene group, were also identified as being important for ligands containing flexible skeletons.

Software packages such as Schrödinger Phase,[17] the HypoGen module of Catalyst[18] and CoMFA[19] can be used to develop 3D QSAR models, which correlate biological activities with 3D structural features of molecules. The HypoGen module of Catalyst generates "hypotheses" (possible pharmacophores) by matching structural features between conformer libraries of active compounds. The hypotheses can then be used to predict the activity of compounds, on the basis of how well the features in compounds match the hypotheses.[20] Phase, on the other hand, generates a hypothesis with a grid-based 3D QSAR method, in which the grid positions of atoms in molecules overlaid to the hypotheses are correlated to their activities using a partial-least-squares (PLS) fitting approach.[17]

One example of the development of downstream inhibitors of NF-κB involved the use of HypoGen to evaluate a library of IKK-β inhibitors to ascertain key features of the molecule class. IKK is a critical enzyme complex in the NF-κB signal transduction cascade, and inhibition of IKK-β causes downregulation of NF-κB.[21] In this study, a library of 159 molecules was developed and 23 of these compounds were chosen as the training set for the QSAR model. The choice of molecule for the training set is critical as the chosen molecules should always be structurally diverse and possess a variety of activities, ranging from low to high. As such, IC_{50} values for the training set ranged from 3 nM to 50 μM, and characteristics such as number of hydrogen-bond acceptors, hydrophobic groups and presence or absence of hydrophobic aromatic groups were used to develop the model. Hypo1 (containing one hydrogen bond donor feature, two ring aromatic features and a hydrophobic moiety) was found to be the best pharmacophore model [characterised by the highest cost difference, the lowest root-mean-square deviation (RMSD), and the best correlation coefficient[22]]. The remaining 136 compounds ('test set') were used to test the model's predictive ability. Interestingly, when the experimentally-derived inhibitory activities were correlated with inhibitory activities as predicted by the pharmacophore model Hypo1, an R^2 of 0.77 was obtained, indicating the validity of this approach in small-molecule drug design.

A similar study used a training set of 35 molecules and determined five key pharmacophore features of the molecule set; three hydrophobic groups, one hydrogen-bond acceptor and one hydrophobic aromatic group (Figure 2.3) using HypoGen. Using a test set of 18 compounds, a correlation of 0.836 was derived between predicted and experimentally derived inhibitory effects.

Furthermore, in this study, docking experiments were performed using the genetic-algorithm-based program GOLD[23] to identify which pocket the derived pharmacophore would be likely to favour. It was found that the hydrophobic aromatic feature was mapped onto the hydrophobic aromatic pocket of the DBR (interacting with amino acids Leu-437 and Val-412). Moreover, the first hydrophobic group (H1) was mapped onto the hydrophobic pocket interacting with Pro-362, the second hydrophobic feature (H2) was mapped onto amino acid Val-358 and the third hydrophobic property (H3) was mapped onto amino acid Cys-359, indicating snug accommodation

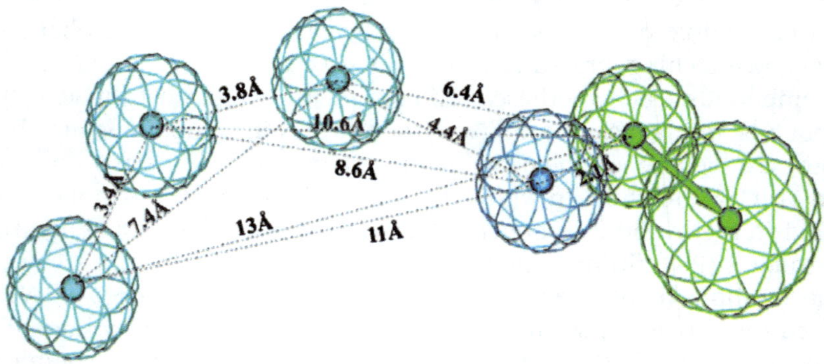

Figure 2.3 The best pharmacophore model of Hypo1. Features are color-coded as follows: hydrophobic aromatic, blue; hydrogen-bond acceptor, green; hydrophobic, light blue.
Reprinted by permission from Springer Nature, *Journal of Molecular Modeling*, S. Nagarajan, A. Ahmed, H. Choo, Y. S. Cho, K-S. Oh, B. H. Lee, K. J. Shin and A. N. Pae, 3D QSAR pharmacophore model based on diverse IKKβ inhibitors, 2010, **17** (2), 209–218, Copyright 2010.[21]

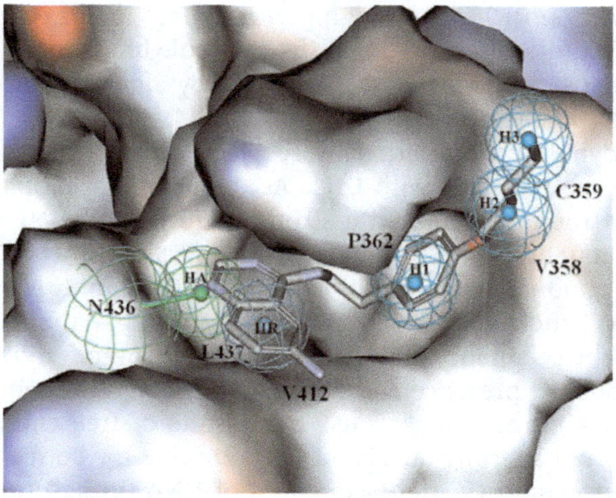

Figure 2.4 Pharmacophore model of Hypo1 and the docking conformation of compound **1** mapped onto the DNA binding region of NF-κB. Features are color-coded as follows: hydrophobic aromatic (HR), blue; hydrogen-bond acceptor (HA), green; hydrophobic (H), light blue.
Reprinted from *Bioorganic & Medicinal Chemistry Letters*, **19** (19), K-C. Tsai, L-W. Teng, Y-M. Shao, Y-C. Chen, Y-C. Lee, M. Li, N-W. Hsiao, The first pharmacophore model for potent NF-κB inhibitors, 5665–5669, Copyright 2009, with permission from Elsevier.[24]

in the DBR. Additionally, the hydrogen bond acceptor feature (HA) formed hydrogen bond interactions with the side chain of amino acid Asn-436,[24] suggesting favourable interaction with the binding pocket (Figure 2.4).

2.2.3 Molecular Dynamics Simulations

Molecular dynamics simulations consist of the calculation of the time-dependent behaviour of molecular systems and have provided valuable information on the changes in conformations of biomolecules (for example proteins or nucleic acids) as predicted over a certain time-course. Although mainly used in the prediction of molecular-level detail of protein structures (*e.g.*, mechanisms of protein folding[25]), the methodologies have recently been adapted to evaluate the interaction of DNA-binding agents with the NF-κB consensus sequence.

The pyrrolobenzodiazepine-based KMR-28-39 alkylates the C2-amino of guanine bases in the minor groove of DNA spanning six base pairs in the process.[26] It has significant *in vitro* cytotoxicity in a number of tumour cell lines including NF-κB-dependent Mia PaCa 2 and MDA-MB-231 (low picomolar). Preliminary studies in a Human Tumour Xenograft mouse model of pancreatic cancer have provided strong evidence of *in vivo* activity at low dose levels (*e.g.*, 300 μg Kg^{-1}) which produce no apparent signs of general toxicity in the mouse. Biophysical evidence [*e.g.*, fluorescence resonance energy transfer (FRET) and high-performance liquid chromatography (HPLC)-based studies] indicated that the molecule binds to the transcription factor consensus sequence of NF-κB, thereby downregulating the transcription factor in the process.

Molecular dynamics simulations were used to investigate the mechanism of binding of the molecule with DNA. As the results of biophysical studies indicated that the molecule had a strong affinity for the NF-κB consensus recognition sequences 5′-GGGGGTCGCC-3′ and 5′-GGGACAGCCC-3′, simulations were conducted on these sequences. This study was undertaken using AMBER11 molecular dynamics software[27] and analysis involved close inspection of the occupancy of the molecule in the minor groove, shape-fit in the binding pocket, presence/absence of H-bonding functional groups and free energy of binding calculations. The results of free energy of binding studies supported a strong affinity of the molecule for both the NF-κB consensus sequences and GC-rich sequences of DNA (in accordance with the results from biophysical experiments), and evaluation of simulations indicated an important role for hydrogen bond acceptors in elucidating sequence-specific binding. The ring nitrogen (N3) of imidazole and the terminal ester of the ligand (both hydrogen bond acceptors) were shown to be critical for binding to the NF-κB consensus sequence, and the distance between these two moieties allowed each to pick up a H-bond with a guanine base.

2.3 STAT3

Virtual screening approaches towards STAT3 are mostly based on the availability of crystallographic data from STAT1 and STAT3 dimers.[28] In a similar manner to NF-κB, dimerization of STAT3 can be targeted in an effort to develop inhibitors of the transcription factor. Turkson and colleagues illustrated that effective disruption of the STAT3–STAT3–DNA complex could be accomplished through the use of a truncated version of STAT3's cognate phosphopeptide binding sequence derived from residues surrounding the pY705 site, PpYLKTK.[29]

The results of computational analysis indicate that the hexapeptide binding site has a number of druggable characteristics, and three distinct binding sites have been identified. Firstly, there are a large number of available hydrogen-bonding interactions, facilitated by the prevalence of polar residues. Secondly, the sub-pocket corresponding to the "LK" portion of the phosphopeptide binding site contains a large hydrophobic region (*i.e.*, T620, K626, Q635 and S636), which is an ideal binding site for hydrophobic ligand fragments, and, finally, the "P" component of the hexapeptide binds to a pocket containing T620, K626, Q635 and S636. These bases are unique to STAT3, and targeting of this particular region can result in selective inhibition of STAT3 over other family members (*e.g.*, STAT1) (Figure 2.5).

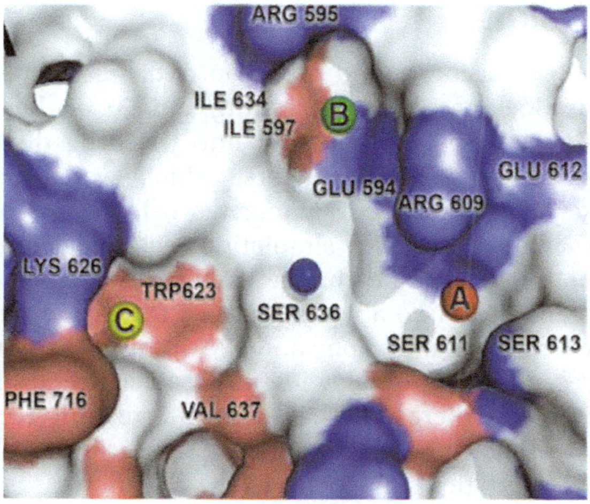

Figure 2.5 Illustration of the key amino acid residues involved in STAT3 binding (PDB ID: 1BG1).[30]
Reprinted from V. M. Shahani, P. Yue, S. Haftchenary, W. Zhao, J. L. Lukkarila, X. Zhang, D. Ball, C. Nona, P. T. Gunning, J. Turkson, *ACS Medicinal Chemistry Letters,* 2011, **2**, 79–84. https://pubs.acs.org/doi/10.1021/ml100224d Further permissions related to the material excerpted should be directed to the ACS.

2.3.1 Docking and Virtual Screening

A number of studies have been reported which incorporate an *in silico* component in the development of STAT3 inhibitors.[28b,31] Szelag and colleagues[32] evaluated the binding potential of inhibitors of the STAT class (*i.e.*, both STAT1 and STAT3). In this study, new 3D structure models for all human (h)STATs were developed and a docking procedure was undertaken to obtain insight into the STAT-SH2 cross-binding specificity of a selection of previously identified STAT3 inhibitors. It was noted that by primarily targeting the highly conserved pTyr-SH2 binding pocket, the majority of the compounds exhibited similar binding affinity and tendency scores for all STATs.

3D structure models for all (h)STATs (1, 2, 3, 4, 5A, 5B and 6) were generated, and a STAT inhibitor screening method was developed based on comparative *in silico* virtual screening and docking validation. This was undertaken to obtain further insight into STAT-SH2 cross-binding specificity of a selection of previously identified STAT3 inhibitors. The standard selection criteria of these compounds were confirmed *in vitro* by the disruption of the phosphotyrosine–SH2 interactions and proven STAT cross-binding. By comparative virtual screening of a natural compound and clean leads library for binding of all STATs, and through the introduction of the 'STAT comparative binding affinity value' (STAT-CBAV) and 'ligand binding pose variation' (LBPV) parameter as selection criteria, initial proof was obtained that indicated that novel *in silico* screening methodologies enable the selection of STAT3-specific inhibitors.

2.4 Conclusion

The improvement in accuracy and precision of *in silico* screening methodologies, such as docking, virtual screening and QSAR, has allowed each of these techniques to be incorporated into the drug discovery process. Proteins, such as transcription factors, which were once considered 'undruggable', can now be effectively targeted using these techniques, and advances in other more complex simulation technologies (*e.g.*, molecular dynamics simulations) indicate that researchers may soon be able to simulate full protein–ligand complexes (derived through experimental techniques such as X-ray crystallography and NMR) in order to optimise ligand contacts with receptors and quickly identify lead molecules for further development. It is likely, therefore, that *in silico* screening technologies will play a pivotal role in both the development of transcription factor inhibitors and the drug discovery field in general in the coming years.

References

1. D. N. Arnosti, *Annu. Rev. Entomol.*, 2003, **48**, 579–602.
2. J. E. Coleman, *Annu. Rev. Biochem.*, 1992, **61**, 897–946.
3. (a) J. E. Darnell Jr., *Nat. Rev. Cancer*, 2002, **2**, 740–749; (b) J. Bromberg, *Breast Cancer Res.*, 2000, **2**, 86–90; (c) J. Bromberg and J. E. Darnell Jr., *Oncogene*, 2000, **19**, 2468–2473; (d) T. Bowman, R. Garcia, J. Turkson and R. Jove, *Oncogene*, 2000, **19**, 2474–2488; (e) J. Turkson and R. Jove, *Oncogene*, 2000, **19**, 6613–6626; (f) J. Turkson, *Expert Opin. Ther. Targets*, 2004, **8**, 409–422; (g) R. Buettner, L. B. Mora and R. Jove, *Clin. Cancer Res.*, 2002, **8**, 945–954; (h) J. E. Darnell, *Nat. Med.*, 2005, **11**, 595–596.
4. K. Shuai, C. M. Horvath, L. H. Huang, S. A. Qureshi, D. Cowburn and J. E. Darnell Jr., *Cell*, 1994, **76**, 821–828.
5. J. Turkson, J. S. Kim, S. Zhang, J. Yuan, M. Huang, M. Glenn, E. Haura, S. Sebti, A. D. Hamilton and R. Jove, *Mol. Cancer Ther.*, 2004, **3**, 261–269.
6. S. Fletcher, J. Turkson and P. T. Gunning, *ChemMedChem*, 2008, **3**, 1159–1168.
7. I. H. Park and C. Li, *J. Mol. Recognit.*, 2011, **24**, 254–265.
8. (a) C. Yanofsky, *Annu. Rev. Biochem.*, 2001, **70**, 1–37; (b) V. Pande, R. K. Sharma, J. Inoue, M. Otsuka and M. J. Ramos, *J. Comput.-Aided Mol. Des.*, 2003, **17**, 825–836.
9. (a) M. Tompa, N. Li, T. L. Bailey, G. M. Church, B. De Moor, E. Eskin, A. V. Favorov, M. C. Frith, Y. Fu, W. J. Kent, V. J. Makeev, A. A. Mironov, W. S. Noble, G. Pavesi, G. Pesole, M. Regnier, N. Simonis, S. Sinha, G. Thijs, J. van Helden, M. Vandenbogaert, Z. Weng, C. Workman, C. Ye and Z. Zhu, *Nat. Biotechnol.*, 2005, **23**, 137–144; (b) Y. Yamamoto and R. B. Gaynor, *J. Clin. Invest.*, 2001, **107**, 135–142; (c) J. Levita, A. A. Nawawi, A. Mutalib and S. Ibrahim, *Int. J. Pharmacol.*, 2010, **6**, 569–576.

10. (a) B. Berkowitz, D. B. Huang, F. E. Chen-Park, P. B. Sigler and G. Ghosh, *J. Biol. Chem.*, 2002, **277**, 24694–24700; (b) F. E. Chen, D. B. Huang, Y. Q. Chen and G. Ghosh, *Nature*, 1998, **391**, 410–413; (c) C. W. Muller, F. A. Rey, M. Sodeoka, G. L. Verdine and S. C. Harrison, *Nature*, 1995, **373**, 311–317; (d) A. S. Prasad, B. Bao, F. W. Beck and F. H. Sarkar, *J. Lab. Clin. Med.*, 2001, **138**, 250–256.
11. (a) J. Mortier, B. Masereel, C. Remouchamps, C. Ganeff, J. Piette and R. Frederick, *Bioorg. Med. Chem. Lett.*, 2010, **20**, 4515–4520; (b) L. Piccagli, M. Borgatti, E. Nicolis, N. Bianchi, I. Mancini, I. Lampronti, D. Vevaldi, F. Dall'Acqua, G. Cabrini and R. Gambari, *Bioorg. Med. Chem.*, 2010, **18**, 8341–8349; (c) L. Piccagli, E. Fabbri, M. Borgatti, V. Bezzerri, I. Mancini, E. Nicolis, M. C. Dechecchi, I. Lampronti, G. Cabrini and R. Gambari, *BMC Struct. Biol.*, 2008, **8**, 38; (d) L. Piccagli, E. Fabbri, M. Borgatti, N. Bianchi, V. Bezzerri, I. Mancini, E. Nicolis, C. M. Dechecchi, I. Lampronti, G. Cabrini and R. Gambari, *ChemMedChem*, 2009, **4**, 2024–2033.
12. W. J. Allen, T. E. Balius, S. Mukherjee, S. R. Brozell, D. T. Moustakas, P. T. Lang, D. A. Case, I. D. Kuntz and R. C. Rizzo, *J. Comput. Chem.*, 2015, **36**, 1132–1156.
13. G. M. Morris, R. Huey, W. Lindstrom, M. F. Sanner, R. K. Belew, D. S. Goodsell and A. J. Olson, *J. Comput. Chem.*, 2009, **30**, 2785–2791.
14. R. A. Friesner, R. B. Murphy, M. P. Repasky, L. L. Frye, J. R. Greenwood, T. A. Halgren, P. C. Sanschagrin and D. T. Mainz, *J. Med. Chem.*, 2006, **49**, 6177–6196.
15. J. J. Irwin and B. K. Shoichet, *J. Chem. Inf. Model.*, 2005, **45**, 177–182.
16. (a) B. Siedle, A. J. García-Piñeres, R. Murillo, J. Schulte-Mönting, V. Castro, P. Rüngeler, C. A. Klaas, F. B. Da Costa, W. Kisiel and I. Merfort, *J. Med. Chem.*, 2004, **47**, 6042–6054; (b) G. Marzaro, A. Guiotto, M. Borgatti, A. Finotti, R. Gambari, G. Breveglieri and A. Chilin, *J. Med. Chem.*, 2013, **56**, 1830–1842.
17. S. L. Dixon, A. M. Smondyrev, E. H. Knoll, S. N. Rao, D. E. Shaw and R. A. Friesner, *J. Comput.-Aided Mol. Des.*, 2006, **20**, 647–671.
18. G. Osman, C. Omoshile and K. Yasuhisa, *Curr. Med. Chem.*, 2004, **11**, 2991–3005.
19. R. D. Cramer, D. E. Patterson and J. D. Bunce, *J. Am. Chem. Soc.*, 1988, **110**, 5959–5967.
20. D. A. Evans, T. N. Doman, D. A. Thorner and M. J. Bodkin, *J. Chem. Inf. Model.*, 2007, **47**, 1248–1257.
21. S. Nagarajan, A. Ahmed, H. Choo, Y. S. Cho, K.-S. Oh, B. H. Lee, K. J. Shin and A. N. Pae, *J. Mol. Model.*, 2011, **17**, 209–218.
22. M. Ravikumar, S. Pavan, S. Bairy, A. B. Pramod, M. Sumakanth, M. Kishore and T. Sumithra, *Chem. Biol. Drug Des.*, 2008, **72**, 79–90.
23. G. Jones, P. Willett, R. C. Glen, A. R. Leach and R. Taylor, *J. Mol. Biol.*, 1997, **267**, 727–748.
24. K.-C. Tsai, L.-W. Teng, Y.-M. Shao, Y.-C. Chen, Y.-C. Lee, M. Li and N.-W. Hsiao, *Bioorg. Med. Chem. Lett.*, 2009, **19**, 5665–5669.

25. I. Daidone, A. Amadei, D. Roccatano and A. D. Nola, *Biophys. J.*, 2003, **85**, 2865–2871.
26. K. M. Rahman, P. J. Jackson, C. H. James, B. P. Basu, J. A. Hartley, M. de la Fuente, A. Schatzlein, M. Robson, R. B. Pedley, C. Pepper, K. R. Fox, P. W. Howard and D. E. Thurston, *J. Med. Chem.*, 2013, **56**, 2911–2935.
27. D. A. Case, Darden, T. A. Cheatham III, T. E. Simmerling, C. L. Wang, J. Duke, R. E. Luo, R. Walker, R. C. Zhang, W. Merz, K. M. Roberts, B. Wang, B. Hayik, S. Roitberg, A. Seabra, G. Kolossváry, I. Wong, K. F. Paesani, F. Vanicek, J. Liu, J. Wu, X. Brozell, S. R. Steinbrecher, T. Gohlke, H. Cai, Q. Ye, X. Wang, J. Hsieh, M.-J. Cui, G. Roe, D. R. Mathews, D. H. Seetin, M. G. Sagui, C. Babin, V. Luchko, T. Gusarov, S. Kovalenko and P. A. Kollman, AMBER 11, University of California, San Francisco 2010.
28. (a) J. S. McMurray, *Biopolymers*, 2008, **90**, 69–79; (b) K. Siddiquee, S. Zhang, W. C. Guida, M. A. Blaskovich, B. Greedy, H. R. Lawrence, M. L. R. Yip, R. Jove, M. M. McLaughlin, N. J. Lawrence, S. M. Sebti and J. Turkson, *Proc. Natl. Acad. Sci. U. S. A.*, 2007, **104**, 7391–7396.
29. J. Turkson, D. Ryan, J. S. Kim, Y. Zhang, Z. Chen, E. Haura, A. Laudano, S. Sebti, A. D. Hamilton and R. Jove, *J. Biol. Chem.*, 2001, **276**, 45443–45455.
30. V. M. Shahani, P. Yue, S. Haftchenary, W. Zhao, J. L. Lukkarila, X. Zhang, D. Ball, C. Nona, P. T. Gunning and J. Turkson, *ACS Med. Chem. Lett.*, 2011, **2**, 79–84.
31. X. L. Zhang, P. B. Yue, B. D. G. Page, T. S. Li, W. Zhao, A. T. Namanja, D. Paladino, J. H. Zhao, Y. Chen, P. T. Gunning and J. Turkson, *Proc. Natl. Acad. Sci. U. S. A.*, 2012, **109**, 9623–9628.
32. M. Szelag, A. Czerwoniec, J. Wesoly and H. A. R. Bluyssen, *PLoS One*, 2015, **10**, e0116688.

CHAPTER 3

Natural Products as Promising Leads Against Oncogenic Transcription Factors and Associated Signalling Pathways

TAUFIQ RAHMAN* AND XAVIER CHEE WEZEN

Department of Pharmacology, University of Cambridge,
Tennis Court Road, Cambridge CB2 1PD, UK
*Email: mtur2@cam.ac.uk

3.1 Introduction

Cancer is a multifactorial heterogeneous disease with abnormalities in various cell signalling pathways implicating oncogenes and tumour suppressor genes, the majority of which encode transcription factors.[1] It is well established[2] by now that deregulated expression or behaviour of a set of transcription factors underlie almost all the hallmarks of cancer described in the much celebrated reviews by Hanahan and Weinberg.[3,4] Mechanistically, transcription factors function by binding to specific DNA response elements, or by forming complexes with other transcription factors, and recruiting chromatin-modifying complexes to their various genetic target loci. Because they mediate their action largely through protein–protein interactions rather than acting like enzymes, they have traditionally been regarded as 'undruggable'.[1] At present, several factors could challenge this notion. Firstly, oncogenic signalling cascades made of diverse upstream

Drug Discovery Series No. 65
Small-molecule Transcription Factor Inhibitors in Oncology
Edited by Khondaker Miraz Rahman and David E. Thurston

Published by the Royal Society of Chemistry, www.rsc.org

components seem to converge only on a handful of transcription factors,[5] thus making the latter attractive points for specific intervention with the possibility of lesser side effects. Secondly, various research groups have come up with strategies including small molecules that could successfully incapacitate transcription factors, disrupting protein–protein or protein–nucleic acid interactions. One glaring proof that transcription factors can indeed be a valid therapeutic target includes the fact that more than 10% of FDA-approved drugs target the nuclear hormone receptors,[6] which are effectively ligand-activated transcription factors.

The role of natural products in providing leads for the development of modern drugs cannot be over-emphasized. Many currently used anti-cancer drugs have a natural origin or are the semi-synthetic derivatives from the natural leads.[7] Growing evidence indicates that natural products can be potent inhibitors of some oncogenic transcription factors. In this review, we seek to illustrate, based on a comprehensive review of the literature, how some naturally occurring small molecules can serve as leads for future development of transcription factor inhibitory drugs. For this, we have chosen few oncogenic transcription factors, namely NF-κB, STAT3, HIF-1 and FOXM1 as example cases.

3.2 Natural Leads Targeting NF-κB

The nuclear factor kappa B (NF-κB) family of proteins [also known as reticuloendotheliosis viral oncogene homolog (Rel) proteins] are ubiquitously-expressed transcription factors, playing key roles in determining cell fate during immune, inflammatory and stress responses. NF-κB controls the expression of over 500 genes involved in cellular transformation, survival, proliferation, invasion, angiogenesis, metastasis and inflammation.[8] It is therefore, unsurprising that deregulated activation of NF-κB is found in many cancers, such as Hodgkin's disease, acute lymphoblastic leukaemia, breast cancer, colon cancer, ovarian cancer, prostate cancer, liver cancer and melanoma.[9]

The NF-κB/Rel family consists of five closely related DNA binding proteins: RelA (p65), RelB, c-Rel, NF-κB1/p50 and NF-κB2/p52, which function as various homodimers and heterodimers with the p50–p65 (RelA) dimer being the most common cellular variant. At their N terminus, all five Rel proteins contain a highly-conserved Rel homology domain (RHD) which is critical for their dimerization, nuclear translocation, DNA binding and interaction with their cytosolic inhibitory proteins (IκBs). In unstimulated cells, NF-κB dimers remain in the cytosol as an inactive complex with the inhibitory IκB proteins. The latter use their ankyrin repeat domains (ARDs) to interact with the RHDs of NF-κB proteins, thereby masking the nuclear localisation sequence and thus preventing the nuclear translocation of these transcription factors. Degradation of IκBs is a prerequisite for NF-κB activation. This is initiated by activation of the IκB kinase (IKK) complex in response to a wide array of stimuli, such as pro-inflammatory cytokines [*e.g.* tumour necrosis factor alpha (TNF-α) or interleukin 1 beta (IL-1β)], oxidative stress, ionizing radiation, microbial toxins, drugs *etc.* IKK phosphorylates specific Ser residues of the IκB protein which is then polyubiquinated and subsequently degraded by the 26S proteasome. The freed NF-κB then translocates into the nucleus, where it activates its target genes, largely through recruiting chromatin-remodelling proteins to their promoters.[8]

Mounting evidence indicates that NF-κB hyperactivity not only enables malignant transformation and tumour progression, but also provides a mechanism by which tumour cells escape immune surveillance and resist therapy. Tumour-promoting inflammation and avoiding immune destruction are notable among the emerging carcinogenic factors[4] and deregulated NF-κB seem to play key role in both, thereby making it a potential drug target in haematological malignancies and solid tumours.[8]

3.2.1 Molecular Targeting of the NF-κB Pathway

Given the well-established role of NF-κB proteins in cancer and other diseases, chemical targeting of the NF-κB pathway has been extensively explored worldwide in past decades and by now over 750 inhibitors of the NF-κB pathway have been reported, including a wide variety of natural and synthetic molecules.[10] Mechanistically, these NF-κB inhibitors can be categorized broadly into directly and indirectly-acting agents. Directly acting inhibitors interact directly with the NF-κB proteins, thus interfering either with their nuclear translocation or DNA binding. Indirectly acting inhibitors of NF-κB signalling are more common and these molecules effectively impair proteosomal degradation of the IκB proteins *via* acting on targets other than the NF-κB proteins that include IκB, IKK, 26S proteosome and proteins that lie upstream of IKK (cell surface receptors, adaptor proteins *etc.*). Several comprehensive reviews are available on natural products as NF-κB inhibitors, acting either directly or directly[8,10–12] and we will therefore mention a sub-set of them to illustrate the principles of their action.

3.2.1.1 Natural Products Inhibiting the Transcriptional Activity of Nuclear NF-κB

There are small molecules, including natural products, that seem to suppress transcriptional activity of the nuclear NF-κB, through either perturbing its DNA and/or recruitment of other regulatory proteins.[8,10,12] Given the large interaction surface mediating the binding of NF-κB to DNA, it is rather hard to expect that the chemical space of natural products would provide suitable, largely non-polar scaffolds that can specifically disrupt such interaction.[13] Nevertheless there are a few small-molecule inhibitors of NF-κB (Figure 3.1) reported in the literature.[10–12] For example, growing evidence indicates that natural compounds (parthenolide, epoxyquinoids, 15-deoxy-D-12,14-prostaglandin J_2) that target Cys-179 of IKKβ also interact with a conserved Cys in

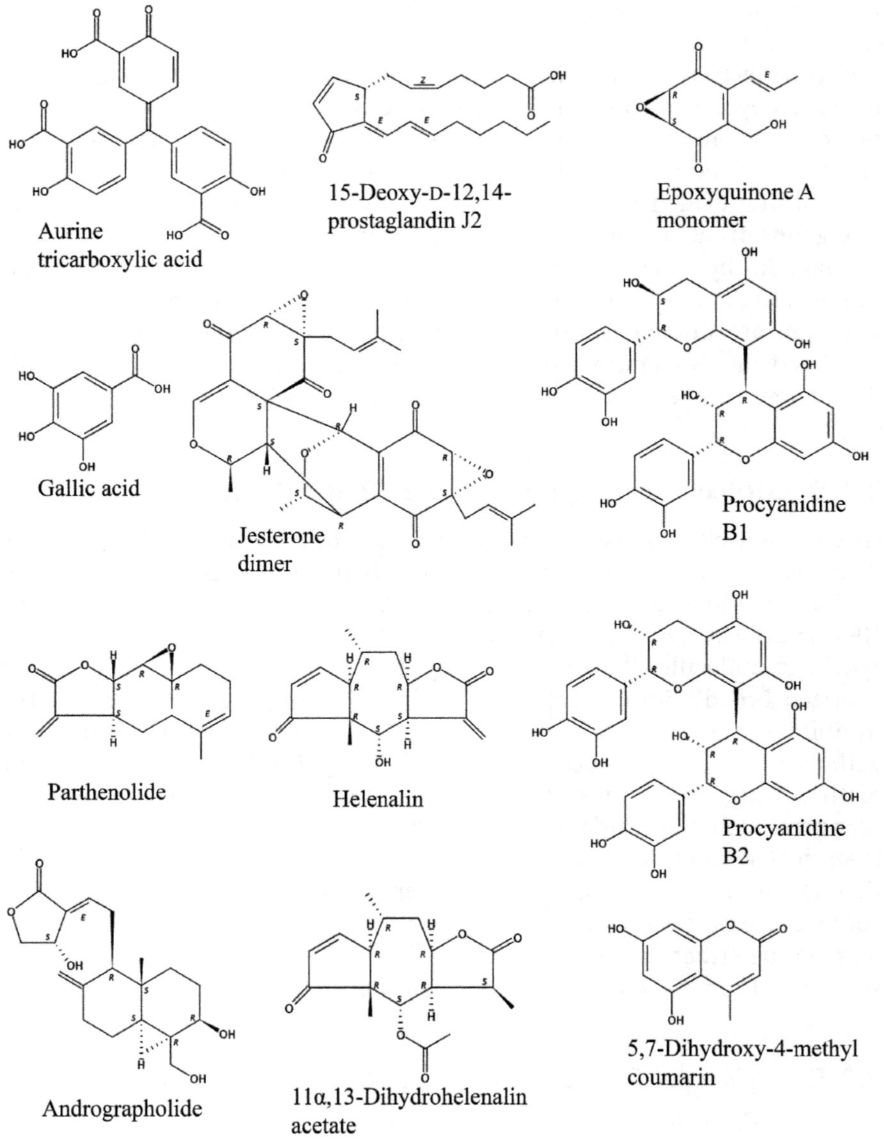

Figure 3.1 Some natural products inhibiting nuclear NF-κB activity.

the DNA-binding loop of NF-κB/Rel proteins to block their DNA-binding ability.[14] A similar Cys-trapping mechanism has been suggested, mainly by computational modelling, for a few other sesquiterpene lactones (helenalin; 11α,13-dihydrohelenalin acetate; 4β,15-epoxy-miller-9*E*-enolide *etc.*) that are thought to interfere with DNA binding of NF-κB.[15–17] Specifically the bioactivity of these lactones has been suggested to stem from a Michael-type addition of the nucleophilic thiol groups Cys 38 and Cys 120 in the p65 monomer of NF-κB to one or more α,β unsaturated carbonyl groups of the compounds.[15–17] The above prediction was somewhat validated through mutagenesis of p65 followed by DNA binding assay.[16,17] Andrographolide, a labdane diterpene isolated from the leaves of *Andrographis paniculata* with known anti-inflammatory and anti-cancer properties, has been shown to form a covalent adduct with Cys 62 of p50.[18] This conjugation, as recently suggested by Nguyen *et al.* (2015)[19] using computational docking and site-directed mutagenesis, seem to be highly dependent on the redox conditions and a few Arg residues in the vicinity of Cys 62 within the conserved RxxRxR motif.

Unlike the Cys modifiers (as Michael acceptors), a few natural poly-hydroxycarboxylated phenolics (*e.g.* gallic acid; aurine tricarboxylic acid; 5,7-dihydroxy-4-methylcoumarin *etc.*) have been shown, *via in silico* docking, to form strong hydrogen bonds with several amino acids, notably Ser 66 of p50 and such an interaction was proposed to sterically hinder the DNA-binding region of p50.[20] Intriguingly, a few other polyphenols, such as B type dimeric procyanidins, have been reported to directly inhibit DNA binding of the p50/RelA and computationally predicted to form hydrogen bonds with Arg 54 and Arg 56 of p50 protein.[21]

3.2.1.2 Natural Products Preventing NF-κB Activation

3.2.1.2.1 Natural Inhibitors of IKK. Among all strategies, inhibiting IKK activity (and thus effectively, IκB degradation) seems to be quite appealing in being the most effective and selective approach for the inhibition of NF-κB activation.[13] Targets other than IκB are as yet unknown for the IKKs and, being an enzyme, it is more druggable. Hitherto over 150 small molecules have been reported as IKK inhibitors, many of which are naturally occurring.[8,10] However, in the majority of the cases, the precise mechanism of action of these chemicals remains unknown. The few IKK inhibitors for which the mechanism of action is known, can be categorized into several classes: (a) ATP-competitive with more selectivity towards IKKs than other kinases (b) allosteric or (c) covalently interacting with a reactive Cys 179 within the activation loop of IKKβ.[8,10] In this review, we will focus only on natural products reported to perturb IKK activity, falling into these categories.

Naturally occurring β-carboline alkaloids and chemicals with β-carboline scaffold are well known to inhibit kinases with variable efficacies, often *via* competing for the ATP-binding site. Castro *et al.* (2003)[22] identified a β-carboline scaffold (5-bromo-6-methoxy-β-carboline) as a weak IKKβ inhibitor, the structural modification of which led to analogues (notably, PS-1145 and ML120B) with significantly improved IKK specificity with potency in nanomolar range.[13,23,24]

Several plant-derived sesquiterpene lactones have been shown to inhibit the activation of NF-κB (Figure 3.2), though their mechanisms of action remain largely unknown.[12] Parthenolide, an abundant sesquiterpene lactone

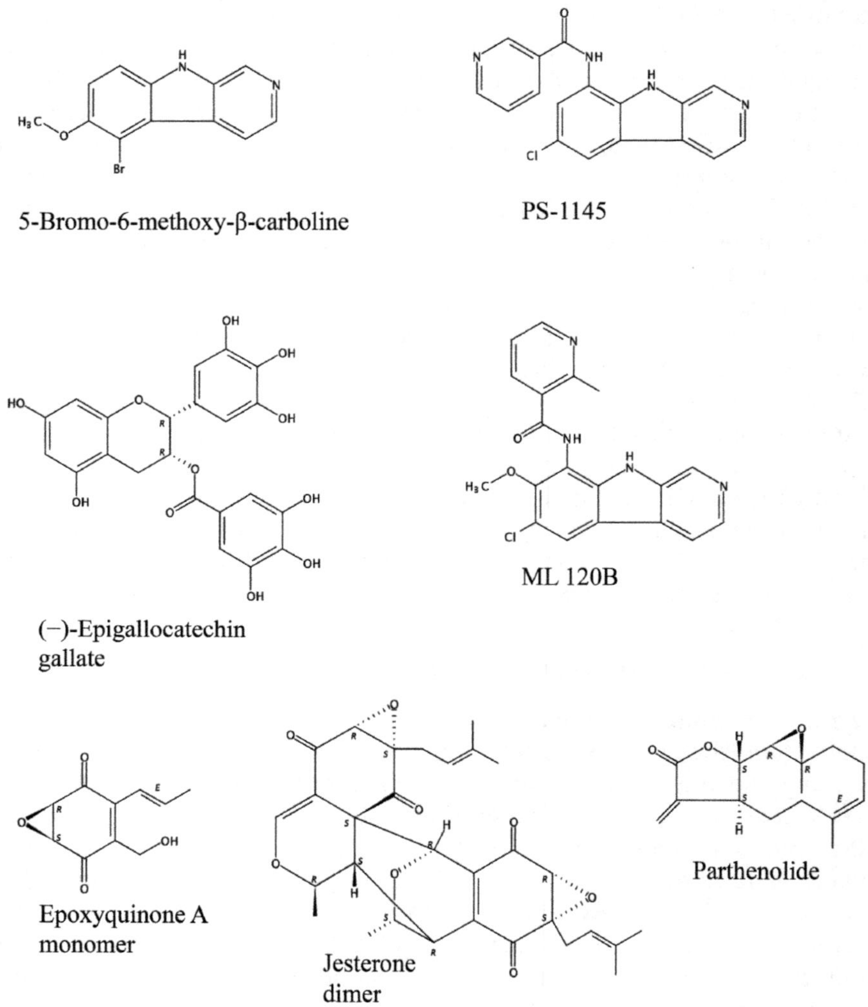

Figure 3.2 A few natural products and derivatives inhibiting NF-κB activation.

found in the medicinal herb Feverfew (*Tanacetum parthenium*), was first reported to inhibit NF-κB activation *via* targeting an unknown signalling component upstream of IκB.[25] Later Kwok *et al.* (2001)[26] using chemical cross linking approaches, revealed that parthenolide interacts with Cys 179 of IKKβ.

In addition to this, epoxyquinone A monomer (EqM), a synthetic derivative of the natural product epoxyquinol A[27] and jesterone dimer, a synthetic dimer of the fungal metabolite jesterone[28] were found to covalently modify Cys 179 of IKKβ and thus eventually inhibit NF-κB activation. These Cys 179-modifying phytochemicals are believed to interfere with phosphorylation-induced IKKβ activation since Cys 179 is located between Ser 177 and Ser

181, which are required for IKKβ activation in response to upstream signals [*e.g.* tumour necrosis factor (TNF), lipopolysaccharide (LPS) *etc.*].[8,10]

Yang *et al.* (2001)[29] reported that the green tea polyphenol, (−)-epigallocatechin gallate (EGCG) which possesses prominent anti-inflammatory and anti-cancer activity, can inhibit the activity of IKK [concentration giving 50% of maximum inhibition (IC_{50}) >18 μM]. The precise mechanism of such inhibition was not studied but the results of limited structure–activity relationship (SAR) studies indicated the gallate moiety to be essential for such role. EGCG appears to be a privileged scaffold, modulating many targets in a variety of pathways. Even in the case of impairing IκB activity, EGCG can also benefit from other mechanisms (proteasome inhibition, discussed below).

Although synthetic non-ATP competitive (*i.e.* allosteric) inhibitors for IKKβ are available (for example, BMS-345541),[13] no such natural NF-κB inhibitor is known so far.

3.2.1.2.2 Natural Inhibitors of IκB Ubiquitination and Proteasomal Degradation. IKK-mediated phosphorylation allows the S-phase kinase-associated protein, Cullin, F-box containing–beta-transducin repeat containing (SCF–β-TrCP) ubiquitin ligase complex to ubiquitinate IκB for its subsequent rapid degradation by the 26S proteasome.[8] Therefore, apart from inhibiting IKK, it is possible to stabilize the inert NF-κB–IκB complex in the cytosol by interfering with the ubiquitin ligase and/or proteasome. As has been reported in some reviews,[8,10] there are many more blockers of the proteasome than of IκB ubiquitination. Among natural blockers of IκB, the *Yersinia* outer protein J (YopJ) protein, a virulence factor for *Yersinia spp.* has been reported to act as a dequbiquitinase for IκB.[30] It is intriguing to note that butyrate and a few other short-chain fatty acids that are produced by the enteric gut flora, can suppress the NF-κB pathway through modulating neddylation of Cullin-1 which is a part of the SCF–β-TrCP ubiquitin ligase complex.[31] While very few synthetic small molecules specifically capable of inhibiting the SCF–β-TrCP ubiquitin ligase complex exist (*e.g.* pyrrolidine dithiocarbamate, PDTC),[32] a naturally occurring small molecule with similar properties is yet to be discovered. It is however worthy of mentioning that two fungal metabolites, namely panepophenanthrin[33] and himeic acid A,[34] have been reported to inhibit the ubiquitin activation enzymes E1 that catalyse the initial step of ubiquitination.[11] Whether these molecules possess NF-κB suppressive ability remains largely unknown.

Many inhibitors of the proteasome have been reported to block NF-κB activation in cell-based systems by blocking proteasome-mediated degradation (but not the phosphorylation) of IκB (Figure 3.3).[10] Bortezomib, a well-established inhibitor of the 26S proteasome, shows promising activity against various human tumours, part of which has been attributed to the interruption caused to NF-κB activation.[8] Among the naturally-occurring proteasomal inhibitors, salinosporamide A (from the marine bacterium

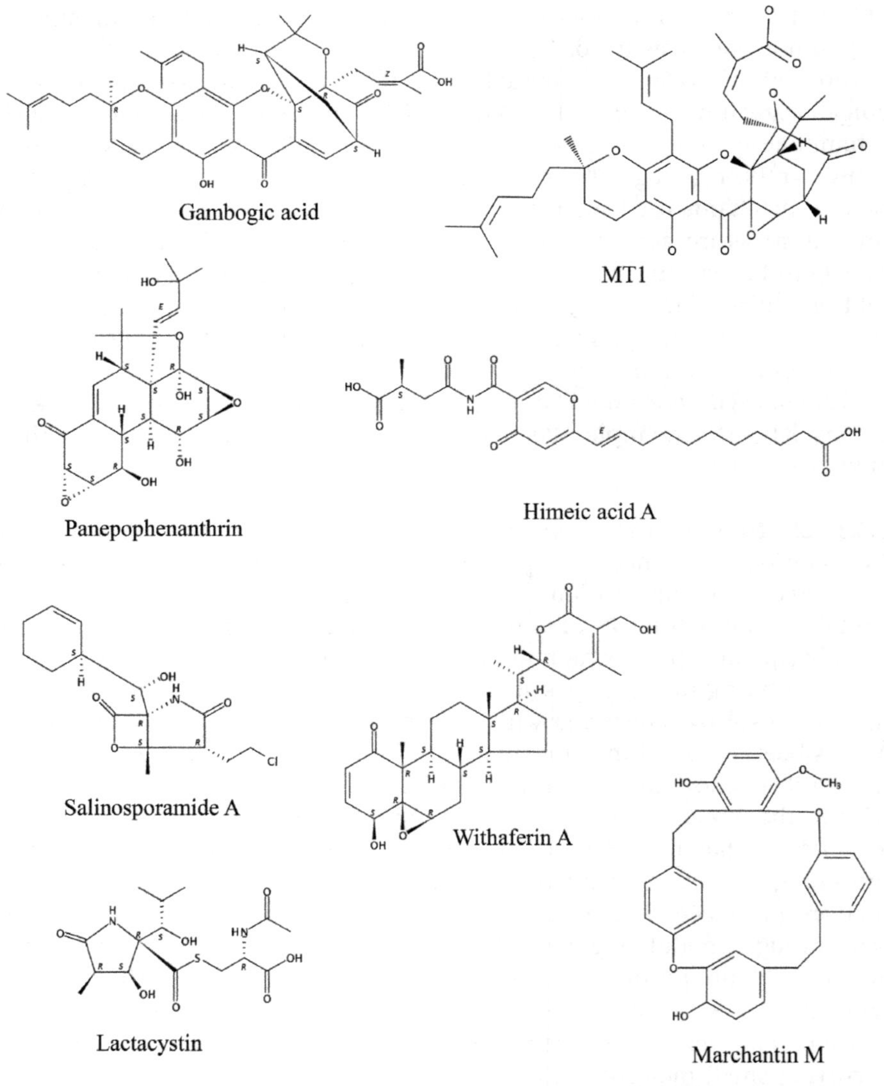

Figure 3.3 A few natural products known to suppress IκB degradation.

Salinispora tropica) exhibits therapeutic potential against a wide variety of tumours through suppressing the NF-κB pathway.[35] The *Streptomyces* product lactacystin (and its synthetic β-lactone precursor) irreversibly block proteasome activity by acylating a Thr residue in the active site of the mammalian proteasome subunit X/MB1[36] and this can perturb nuclear translocation of NF-κB.[37]

Plant polyphenols, such as EGCG, quercetin, chrysin, resveratrol, curcumin, geniestein, oleuropain, 5,6,3′,4′-tetrahydroxy-7-methoxyflavone *etc.* have been suggested to inhibit mammalian proteasomes.[38] The exact mechanism of their action remains largely unexplored, except for EGCG. The latter has been proposed, through *in silico* docking, to bind to the chymotrypsin site in an orientation and conformation that is suitable for a nucleophilic attack by Thr-1.[39] Some limited SAR data is available for these plant phenolics as proteasomal inhibitors,[40] most of these compounds are also known to suppress NF-κB activity, with suggested involvement at multiple steps.[12]

Gambogic acid, (GA), the major active ingredient of resinous exudate from the *Garcinia hanburryi* tree, has been authorised for phase II clinical trials against some tumours in China.[41] Emerging evidence indicates that GA can perturb several oncogenic signalling pathways, including that of NF-κB, transferrin receptor and vascular endothelial growth factor (VEGF) receptor.

Recently, Liu and colleagues (2013)[42] have shown that GA is metabolized by CYP2E1 to produce an epoxide ('MT1') that inhibits the proteasome activity through a possible interaction with the inhibitor of the proteasomal β5 chymotryptic subunit. This is appealing since the expression of the CYP2E1 gene is very high in tumours but low in many normal tissues, GA could therefore produce tissue-specific proteasome inhibition and tumour-specific toxicity. One of the withanolides, namely withaferin A isolated from the Ayurvedic medicinal herb, *Withania somnifera* is known to possess prominent anti-cancer activity and has recently been shown to bind and inhibit the catalytic β subunit of the 20S proteasome specifically.[43] Marchantin M (Mar) is a naturally occurring macrocyclic bisbibenzyl, known to be cytotoxic to human prostate cancer cells. Recently it was shown to inhibit chymotrypsin-like activities of the 20S proteasome and computational modelling indicated that Mar non-covalently binds active sites of proteasome β5 and β1 subunits, resulting in a non-competitive inhibition.[44]

3.3 Natural Leads Targeting STAT3

Signal transducer and activator of transcription 3 (STAT3) belongs to a family of latent cytoplasmic transcription factors that become activated in response to stimulation of cell-surface receptors by cytokines and growth factors. STAT3 is often constitutively active in many human cancer cells, including multiple myeloma, leukaemia, lymphoma and solid tumours, and has been implicated in several hallmarks of cancer, including increased cellular proliferation, resistance to apoptosis, increased angiogenesis and suppression of antitumor immune responses.[45]

STAT molecules exist in a monomeric, latent state in the cytoplasm. Activation of STAT3 (and other STAT proteins) is initiated by phosphorylation of a specific tyrosine residue (Tyr 705) within the transcriptional activation domain (TAD) located towards its C-terminus. This then allows two STAT monomers to dimerise *via* a reciprocal interaction between the phosphorylated Tyr 705 and the Rous sarcoma virus oncogene homologue (Src)-homology 2 (SH2) domain of each monomer. The activated dimer then translocates to the nucleus, where it binds to specific DNA response elements in the promoter regions of target genes (notably those involved in anti-apoptotic, pro-proliferative and immune responses) to regulate their transcription. In addition, phosphorylation of Ser 727 in the TAD of STAT3 seems to be important for exhibiting its maximal transcriptional activity.[46] Several kinases are responsible for the Tyr phosphorylation of STATs, including non-receptor tyrosine kinases [non-RTKs, which upon activation, recruit cytosolic tyrosine kinases like Janus kinases (JAKs) and Src] and receptor tyrosine kinases (RTKs, such as various growth factor receptors). Given the established role of aberrant STAT3 activity in various tumours, development of STAT3-specific inhibitors has been actively pursued in academia as well as industry. Strategies aimed at inhibiting any STAT, including STAT3, could broadly include intervention at (a) targets (RTKs/non-RTKs, cytosolic kinases like JAKs and Src *etc.*) lying upstream of STATs and thus impairing its phosphorylation *en route* to its dimerization and (b) STAT protein itself, thereby preventing its dimerization, nuclear translocation and DNA binding ability. So far, various small molecules, biologics, anti-sense and decoy oligonucleotides have been used for some of these intervention strategies against STAT3.[47] Over the last few decades, many phytochemicals have shown potential to inhibit STAT signalling through different mechanisms. In this review, we have focused on the STAT3-inhibitory phytochemicals that appear to be better-characterised, mechanistically. Also we have excluded those with reported inhibition of the activity and/or expression of the RTKs or non-RTKs and their physiological peptide ligands since those are likely to trigger other cellular effects.

3.3.1 Natural Inhibitors of STAT3 Phosphorylation

Several cucurbitacins (hitherto designated as Cucs). which are tetracyclic triterpenoids initially found in the plants of the Cucurbitaceae family, have been reported to be specific inhibitors of the JAK/STAT signalling without affecting Src, Akt, extracellular-signal-regulated kinase (Erk)1/2, and c-Jun N-terminal kinase (JNK).[48,49] Almost all of them (Cucs B, E, I and Q) appear to significantly reduce the level of Tyr 705-phosphorylated STAT3 in various cancerous cells, without affecting the normal STAT3 level, indicating their likely intervention at the phosphorylation (and thus activation) step (Figure 3.4). Indeed for some STAT3-inhibitory cucurbitacins such as Cuc B,[50] Cuc I (JSI-124)[51] and Cuc E,[52] accompanying inhibition of Tyr phosphorylation (and thus activation) of JAK2 was also noted. Intriguingly, Cuc Q induced apoptosis in tumours through its inhibition of STAT3 activation whilst hardly affecting JAK2.[49] Conversely, Cuc A, though structurally very similar to Cuc Q, inhibited JAK2 but not STAT3 activation and was not able to induce apoptosis and inhibit tumour growth of the A549 lung tumours in nude mice. This indicated that STAT3 inhibition is critical for showing anti-tumour activity. The limited SAR of the tested Cucs thus revealed that addition of a single hydroxyl group to carbon 11 of the Cucs results in loss of anti-STAT3 activity, whereas a simple conversion of a carbon 3 carbonyl to a hydroxyl leads to loss of anti-JAK2 activity.

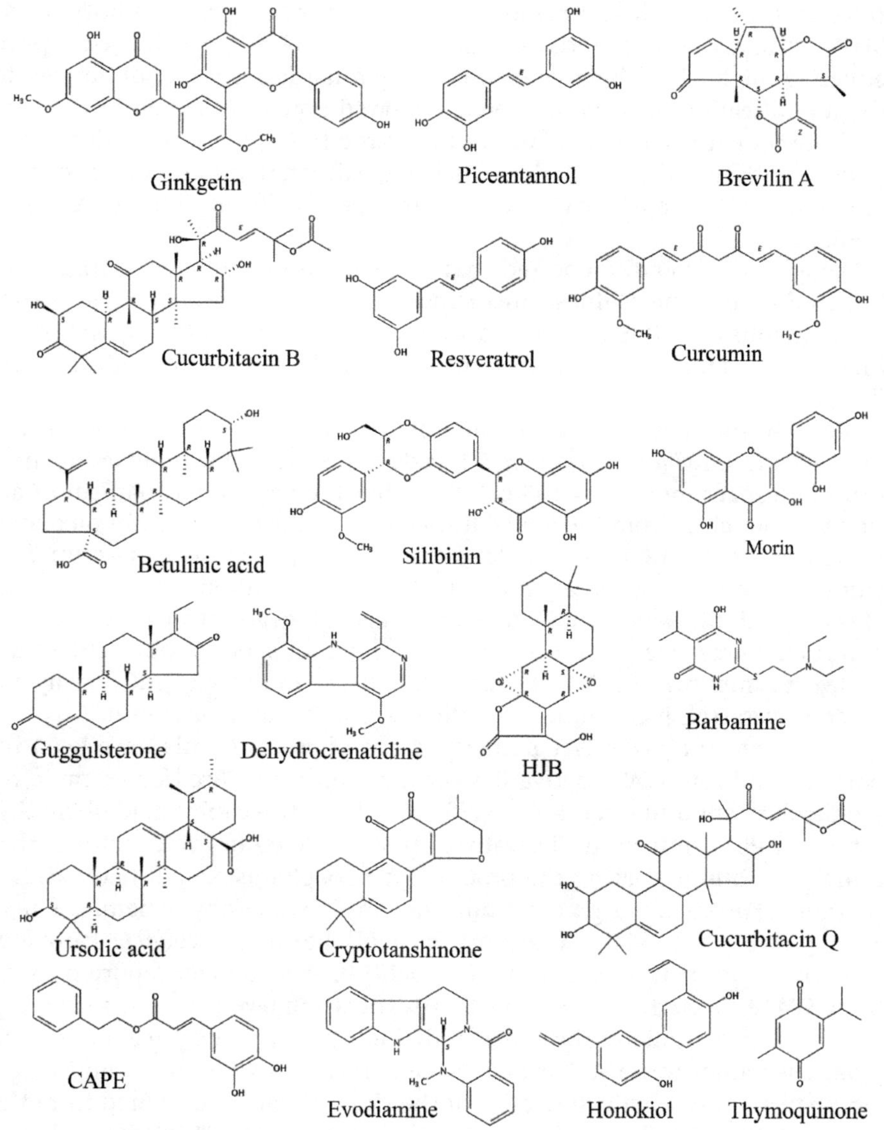

Figure 3.4 Some natural inhibitors of STAT3 phosphorylation.

Cryptotanshinone—a major constituent of the Chinese medicinal herb, *Salvia miltiorrhiza* Bunge (Danshen) was reported to inhibit Tyr 705 phosphorylation of STAT3 without affecting JAK2 or other relevant upstream proteins,[53] suggesting a possible direct effect. Ginkgetin—a biflavonoid from *Ginko biloba*, has been shown to inhibit Tyr 705 phosphorylation of STAT3 through partial inhibition of JAK2 and Src.[54] Guggulsterone [4,17(20)-pregnadiene-3,16-dione; GS], derived from the organic extract of the gum resin from the Guggul tree, inhibited STAT3 activation through inhibition of

JAK2 and Src as well as upregulation of the protein tyrosine phosphatase SHP-1. The action was stereospecific, with the (Z)-GS being the active principle.[55] A high-throughput screening using a natural product library led to the identification of Brevilin A, which showed potent STAT3 and STAT1 inhibitory property *via* specific inhibition of the JAK1 kinase (JH1) domain.[56] Very recently, another natural product—dehydrocrenatidine—was shown to attenuate STAT3 and STAT1 activity by specifically inhibiting JAK2-JH1 domain.[57]

Wang *et al.* (2009)[58] reported that 17-hydroxy-jolkinolide B (HJB), a diterpenoid from the Chinese medicinal herb *Euphorbia fischeriana* Steud, strongly inhibited IL-6-induced as well as constitutive STAT3 activation by inactivating several members of the JAK family through chemical cross-linking.

Silibinin (or silybin), a flavanone obtained from the seed extract of the milk thistle (*Silybum marianum* L.), reduced STAT3 phosphorylation in a prostate cancer model (DU145 cell line), but the precise mechanism of action was not clear from the study. It was noteworthy that silibinin-mediated inhibition of STAT3 phosphorylation led to apoptosis only in serum-free culture conditions, whereas under 10% serum conditions, the anti-STAT3 effect of silibinin was not adequate to trigger cell death, which required the additional presence of a JAK inhibitor.[59] Various other phytochemicals known to suppress STAT3 phosphorylation *via* inhibiting JAK activity include resveratrol, piceatannol, emodin, barbamine and curcumin.[47]

Apart from the JAK–STAT pathway, a few phytochemicals have been reported to inhibit STAT3 activation through inhibiting Src [for example resveratrol; thymoquinone; caffeic acid or its derivative caffeic acid phenethyl ester (CAPE); indirubin derivatives (E804; oleanolic acid and barbamine)][60–62] and activating the protein Tyr-phosphatase small heterodimer partner (SHP)1/2 (for *e.g.*, curcumin, morin).[63] Some phytochemicals (*e.g.* evodiamine, honokiol, 1′-acetoxychavicol acetate, acetyl-11-ketobeta-boswellic acid, betulinic acid, ursolic acid) have also been reported to inhibit STAT3 through upregulating SHP1expression level.[47,48] It is intriguing to note that two microtubule destabilizing agents, namely paclitaxel (originally isolated from the bark of the yew tree, *Taxus brevifolia*) and vinorelbine (a semi-synthetic derivative of the vinca alkaloids obtained from the leaves of *Catharanthus roseus*), seem also to suppress STAT3 phosphorylation in several cancer cells. Of the two, paclitaxel has been shown to disrupt interaction of STAT3 with tubulin.[64] The link between tubulin destabilisation and inhibition of STAT3 phosphorylation, however, remains unclear.

3.3.2 Natural Inhibitors of STAT3 Dimerization

Given the critical role of the SH2 domain in STAT3 dimerization preceded by its phosphorylation, it has been among the highly sought-after targets so far using small molecules as well as peptidomimetic approaches. However, any

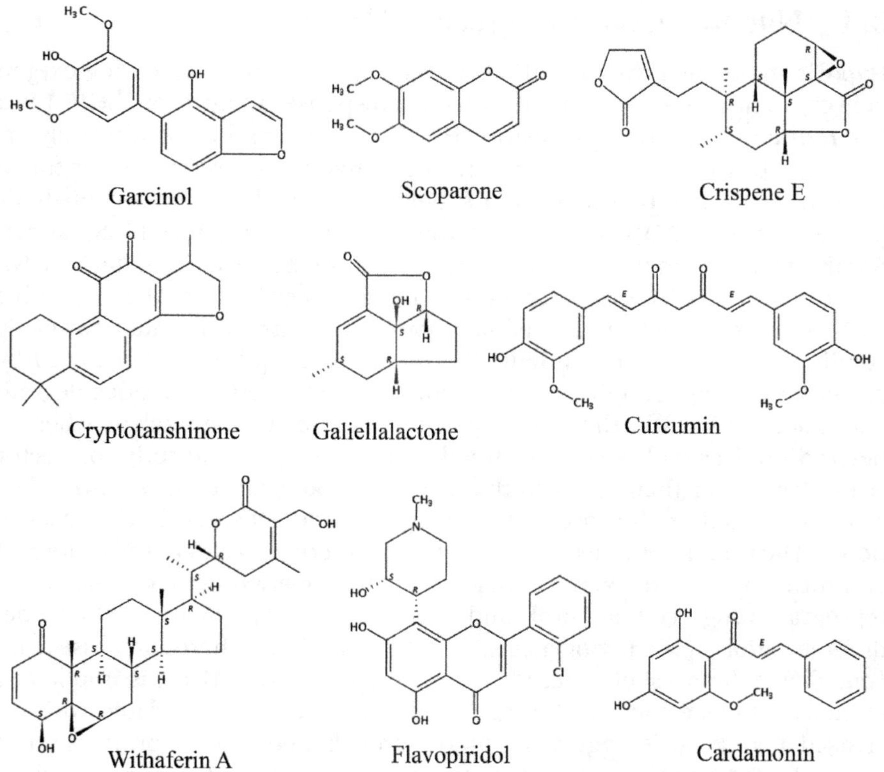

Figure 3.5 A few natural inhibitors of STAT3 dimerization and DNA binding.

naturally occurring anti-STAT3 chemical with experimentally-proven interaction with its SH2 domain is yet to be identified. Recently, there have been reports of several phytochemicals inhibiting STAT3 dimerization (Figure 3.5) and these notably include cryptotanshinone,[53] curcumin and its analogues,[65] garcinol,[66] scoparone,[67] withaferin A,[68] cardamonin[69] and crispene E.[70] The underlying mechanism for all these compounds has been suggested, through *in silico* docking, to be a direct interaction with the SH2 domain of STAT3. This is yet to be validated by suitable wet experiments.

3.3.3 Natural Inhibitors of STAT3–DNA Interaction

Galiellalactone (Figure 3.5) obtained from the fermentations of the ascomycete strain A111-95, has been reported to specifically block, at submicromolar concentrations, the DNA binding of the activated STAT3 dimers whilst not affecting their phosphorylation.[71] Flavopiridol (Figure 3.5), a naturally-occurring flavonoid with known anti-cancer properties, was also found to disrupt the STAT3 dimer–DNA interaction in a DNA cellulose affinity chromatography and nuclear elution assay.[72]

3.4 Natural Leads Targeting HIF-1

Hypoxia-inducible factor-1 (HIF-1) is critically involved in regulating oxygen delivery to tissues and mediating adaptive responses to hypoxia.[73] HIF-1 is a heterodimer of an oxygen-sensitive HIF-1α subunit and a constitutively expressed HIF-1β subunit (also known as the aryl hydrocarbon nuclear translocator, ARNT). Both the subunits belong to the basic helix–loop–helix (bHLH) period (PER)–ARNT–single-minded homolog (SIM) (PAS) superfamily of eukaryotic transcription factors that use the basic domains for DNA binding and the bHLH domain for dimerization. The levels of HIF-1α, unlike HIF-1β, are regulated by intracellular oxygen concentration. Under normoxic conditions, HIF-1α is continually degraded *via* ubiquitination followed by proteosomal degradation. However, during hypoxia, HIF-1α evades degradation, accumulates in the cytosol and translocates to the nucleus where it heterodimerizes with HIF-1β to finally form the active transcription factor HIF-1. The latter then binds to the hypoxia-responsive element (HRE, 5′-G/ACGTG-3′ motif) within over 100 target genes and activates their transcription.[74] The protein products of these genes play crucial roles in the acute and chronic adaptation to oxygen deficiency, including erythropoiesis, metabolic reprogramming, cell survival and inhibition of apoptosis, cellular de-differentiation and angiogenesis.[73,74] Almost all of these processes are *bona fide* 'hallmarks' of cancer,[3,4] it is unsurprising that HIF-1 is implicated in cancer. Indeed increased levels of HIF-1α have been correlated with increased angiogenesis, aggressive tumour growth, poor patient prognosis and reduced effectiveness of radiation and chemotherapy.[75] In animal models, upregulation of HIF-1 leads to increased tumour growth, vascularization and metastasis, whereas its loss of function or downregulation produces the opposite effect, thus validating HIF-1 as a potential anti-cancer target.[73]

3.4.1 Natural Leads Promoting HIF-1α Degradation

Under normoxic conditions, degradation of HIF-1α is initiated by a family of prolyl hydroxylases (PHDs) that uses O_2, iron and 2-oxoglutarate to hydroxylate HIF-1α at Pro 564 and Pro 402. Such hydroxylation allows HIF-1α to bind to the tumour suppressor von Hippel–Lindau protein (pVHL), which also binds to Elongin C and thus recruits a ubiquitin ligase complex that targets HIF-1α for ubiquitination and proteasomal degradation.[73]

Several phytochemicals with anti-tumour and anti-angiogenic activities have been suggested to promote proteosomal degradation of HIF-1α (Figure 3.6). For example, bavachinin,[76] icariside II,[77] rhapontigenin[78] and wondonin[79] have been suggested to enhance the interaction between pVHL and HIF-1α. Activation of PPARα seem to enhance the binding of HIF-1α to pVHL.[80] There are many naturally-occurring PPARα agonists that could potentially promote proteosomal degradation of HIF-1α.

HIF-1α is also known to bind with the molecular chaperone heat shock protein (Hsp)90 *via* the PAS B domain and such association seems to

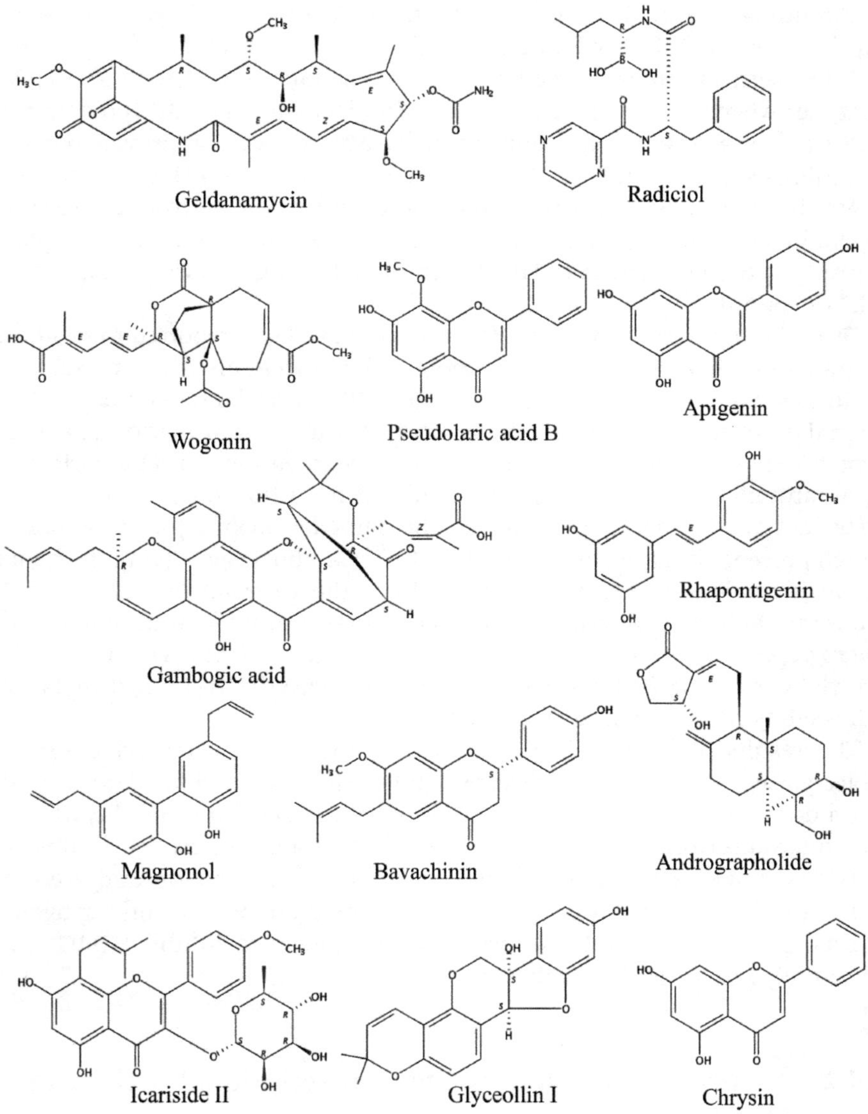

Figure 3.6 Some natural products promoting HIF-1α degradation.

contribute towards HIF-1α stability.[81–83] Known inhibitors of Hsp90 have been reported to promote pVHL-independent degradation of HIF-1α and also to hamper transcriptional activity of HIF-1.[81,84] By now, several natural products are well established as Hsp90 inhibitors.[85] It will be unsurprising if some of these Hsp90 inhibitory anti-tumour natural products also benefit from downregulating the HIF-1 pathway. For example, the natural benzoquinone ansamycin antibiotic geldanamycin promotes proteosomal degradation of HIF-1α.[86] Geldanamycin is known to compete for the

ATP-binding site on Hsp90.[85] A few other phytochemicals such as wogonin,[87] apigenin,[88] chrysin[89] and glyceollins[90] have been reported to promote HIF-1α degradation by impeding the interaction of HIF-1α and Hsp90. Whether wogonin and apigenin directly interact with the PAS B domain of HIF-1α or the ATP-binding domain of Hsp90 remains unproven but the glyceollins seem to bind tightly to the ATP binding site on Hsp90 *in silico.*[90]

Another naturally occurring resorcinol lactone, radiciol, mimics geldanamycin in terms of binding to the ATP site of Hsp90 but unlike geldanamycin, it prevents the interaction of HIF-1 with the HRE of genes without affecting HIF-1α stability.[91]

Several other natural products (Figure 3.6) have been reported to promote degradation of HIF-1α through various other mechanisms. For example, c-Jun has been reported to associate with HIF-1α *via* its oxygen-dependent degradation domain and this masks the sites for ubiquitination and thus protects HIF-1α from proteasome-executing degradation.[92] Pseudolaric acid B, a naturally occurring diterpenoid isolated from the root bark of *Pseudolarix kaempferi* (Pinaceae) enhanced phosphorylation of c-Jun at Ser63/73, which prevents its association with HIF-1α.[93] Gambogic acid, (GA), the major active ingredient of resinous exudate from the *Garcinia hanburryi* tree,[41] andrographolide, a diterpenoid lactone isolated from a traditional herb *Andrographis paniculata*,[94] and magnonol[95] upregulated HIF-1α-specific prolyl hydroxylase (PHD2), which caused increased HIF-1α hydroxylation followed by proteosomal degradation.

The results of biochemical studies have indicated that class I histone deacetylases (HDAC1 and 3) directly bind to the oxygen-dependent degradation domain of HIF-1α and such association increases the stability of HIF-1α and transcriptional activity of HIF-1.[96] Histone deacetylase inhibitors (HDACIs) produce a marked inhibition of HIF-1α expression and are currently in clinical trials, partly on the basis of their potent antiangiogenic effects. Therefore natural products that are known to inhibit HDACs can potentially reduce HIF-1α stability.

3.4.2 Natural Leads Suppressing Transcriptional Activity of HIF-1

The transcriptional activity of HIF-1, in theory, could be inhibited by preventing (a) heterodimerization of HIF-1α and HIF-1β (b) interaction of HIF-1 with the HRE of target gene promoters and (c) interaction of HIF-1 with other co-regulators such as p300/CBP (cyclic adenosine monophosphate-response element-binding protein-binding protein).

Using a HIF-1–targeted cell-based high-throughput screen of the National Cancer Institute 140 000 small-molecule library, Melillo and colleagues (2005)[97] identified echinomycin (NSC-13502), which in nanomolar concentrations interrupted the binding of HIF-1 with the HRE sequence within the promoters of hypoxia-sensitive genes without affecting the stability of

HIF-1α. Echinomycin is a cyclic peptide of the family of quinoxaline antibiotics that was originally isolated from *Streptomyces echinatus.*[98]

For maximal transcriptional activity of HIF-1, it requires interaction with the cofactor CBP/p300 and structural studies have established that the C-terminal activation domain (C-TAD) of HIF-1a directly interacts with the CH1 domain of CBP/p300. Chetomin, an epidithiodiketopiperazine (ETP) metabolite of the fungus *Chaetomium* species (Figure 3.7), has proven to be effective at inhibiting HIF-1α activity by disrupting the binding of p300 to both HIF-1α and HIF-2α, and in attenuating tumour growth in an in

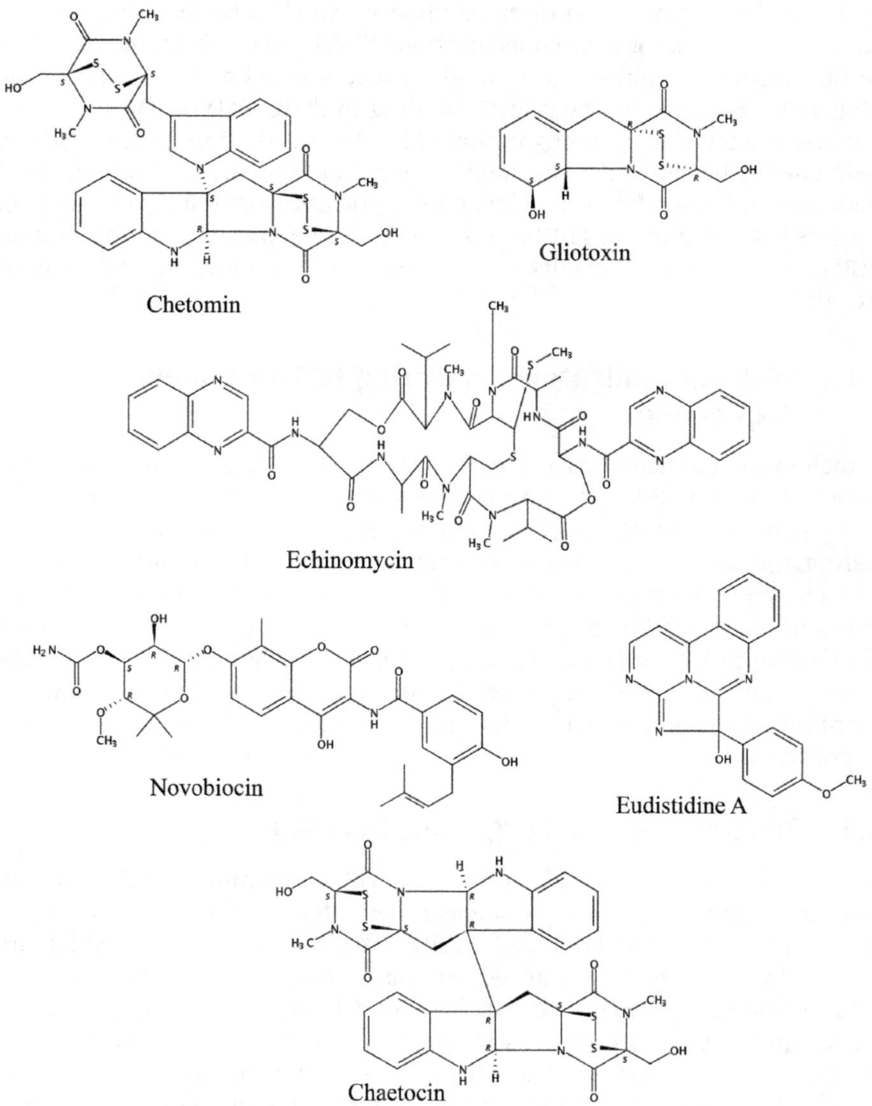

Figure 3.7 Some natural products suppressing HIF-1 transcriptional activity.

HCT116 and PC3 xenograft models.[99] Subsequent work from the Schofield group[100] established that chetomin and other naturally occurring ETPs, such as gliotoxin and chaetocin, work, at least in part, by chelating a structural Zn^{2+} from the CH1 domain of p300, thus destabilising its structure, ablating the interaction with HIF-1α.

Very recently, Hamilton and colleagues (2015)[101] ran a high throughput screening of 10 000 natural-products-based structures and identified two electrophilic chemical classes namely, quinones and indandiones (Figure 3.7) that could potentially inhibit HIF-1α–p300 interaction. Bioassay-guided fractionation of the crude extract of the marine ascidian *Eudistoma* sp led to the identification of eudistidine A, which inhibited p300–HIF-1α interactions at micromolar concentrations.[102] The actinobacterial metabolite novobiocin, an aminocoumarin glycoside, is also known to disrupt the p300–HIF-1α interaction by directly binding to the C-TAD of HIF-1α.[103]

Water extract of red ginseng inhibited HIF1α and 1β dimerization with no toxic effects, however, the mechanism of action and the anti-tumoural effects are not known.[104] Acriflavine, a synthetic antibacterial agent, binds to the PAS-B subdomain of HIF1α and HIF2α thereby prevents the binding to HIF1β, an effect that results in reduced VEGF production and tumour growth.[105]

3.4.3 Natural Leads Downregulating HIF-1α Protein Expression

Mitochondria-mediated reactive oxygen species (ROS) signalling leads to the stabilization of HIF-1α protein and the subsequent activation of HIF-1.[106] Many natural products of plant and marine origin have been shown to disrupt the mitochondrial electron transport chain (ETC) and thus to selectively suppress hypoxia-induced HIF-1 activation.[107] Lot of these molecules have emerged through the HIF-1 inhibitory program at the University of Mississippi led by David Nagle's group. We, however, refrain from discussing these molecules at greater length here and would like to draw the attention of the interested readers to a recent review[107] as well as other recent works.

3.5 Natural Leads Targeting FOXM1

Forkhead box M1 [FOXM1, also known as Trident, membrane palmitoylated protein 2 (MPP2) or hepatocyte nuclear factor–forkhead homologue (HFH)-11] is a member of the Forkhead family of transcription factors, which are defined by the presence of an evolutionarily conserved Forkhead/winged-helix DNA-binding domain (DBD).[108] FOXM1 is expressed in all embryonic tissues and in proliferating cells of epithelial and mesenchymal origin, but its expression is down-regulated in quiescent or terminally differentiated cells.[109] Activation of FOXM1 (and related family members) results in transcriptional induction of relevant genes, including cyclin B, survivin, aurora B kinase, cell division cycle (cdc)25b phosphatase, polo-like kinase 1 (Plk1) and some others. In addition, FOXM1 transcriptionally activates S-phase kinase-associated protein 2 (Skp2) and cyclin-dependent kinases regulatory subunit 1 (Cks1) genes (subunits of the Skp1–Cullin1–F-box ubiquitin ligase complex), thus targeting cyclin dependent kinase (CDK)/cell cycle inhibitors p21 wild type p53-activated fragment (WAF1) and p27kinase inhibitor protein 1 (KIP1) for degradation.[110] All these indicate that the FOXM1 transcription factors are crucial for growth phase 1–is phase (G1–S) and growth phase 2–mitosis phase (G2–M) cell cycle phase progression and mitotic spindle integrity. It is also noteworthy that FOXM1 could be activated by upstream oncogenic Rest sarcoma viral oncogene homologue (Ras)–mitogen activated protein kinase (MAPK), Sonic Hedgehog, NF-κB and epidermal growth factor receptor (EGFR) pathways and is negatively regulated by tumour suppressor p53.[111] It is therefore unsurprising that aberrant activity of FOXM1 and related proteins can predispose to tumourigenesis and cancer progression. FOXM1 indeed is one of the most overexpressed genes in human solid tumours and it is upregulated in various forms of human cancers.[111] Over the years it has become apparent that FOXM1 drives tumorigenesis not only by increasing proliferation but by affecting several 'hallmarks of cancer'[3,4] development and progression, including evasion of the action of tumour suppressors, increasing resistance to apoptosis, imparting replicative immortality and triggering angiogenesis and metastasis. The important role of FOXM1 in cancer affirms its significance for therapeutic intervention. Downregulation of FOXM1 retarded growth and metastasis in various *ex vivo* and *in vivo* models.[111] The inhibition of FOXM1 activity is emerging as an attractive goal for cancer therapy.

3.5.1 Targeting FOXM1 Using Small Molecules

Several anti-cancer drugs and chemopreventive compounds have been reported to downregulate FOXM1 at both the mRNA and protein levels (Figure 3.8). However their precise mechanism of action remains unclear to date. Among these, two natural products or nature-derived molecules are notable–3,3′-diindolylmethane (DIM) and genistein. DIM is a derivative of indole-3-carbinol which is found in cruciferous vegetables, such as broccoli, Brussels sprouts, cabbage and kale.[112] Treatment of breast cancer cells with DIM downregulated FOXM1 mRNA and protein expression. Inhibition of FOXM1 by DIM correlated with decreased cell growth and invasion of breast cancer cells *in vitro*.

In addition, DIM treatment suppressed xenograft tumour growth and FOXM1 expression *in vivo*.[113] Genistein–a natural isoflavonoid found in soybean–and its synthetic analogue, 7-difluoromethoxyl-5,4′-di-n-octyl genistein have been reported to cause significant reduction in FOXM1 mRNA and protein expression and to retard growth in pancreatic, gastric, ovarian and prostate cancer cells.[114–117]

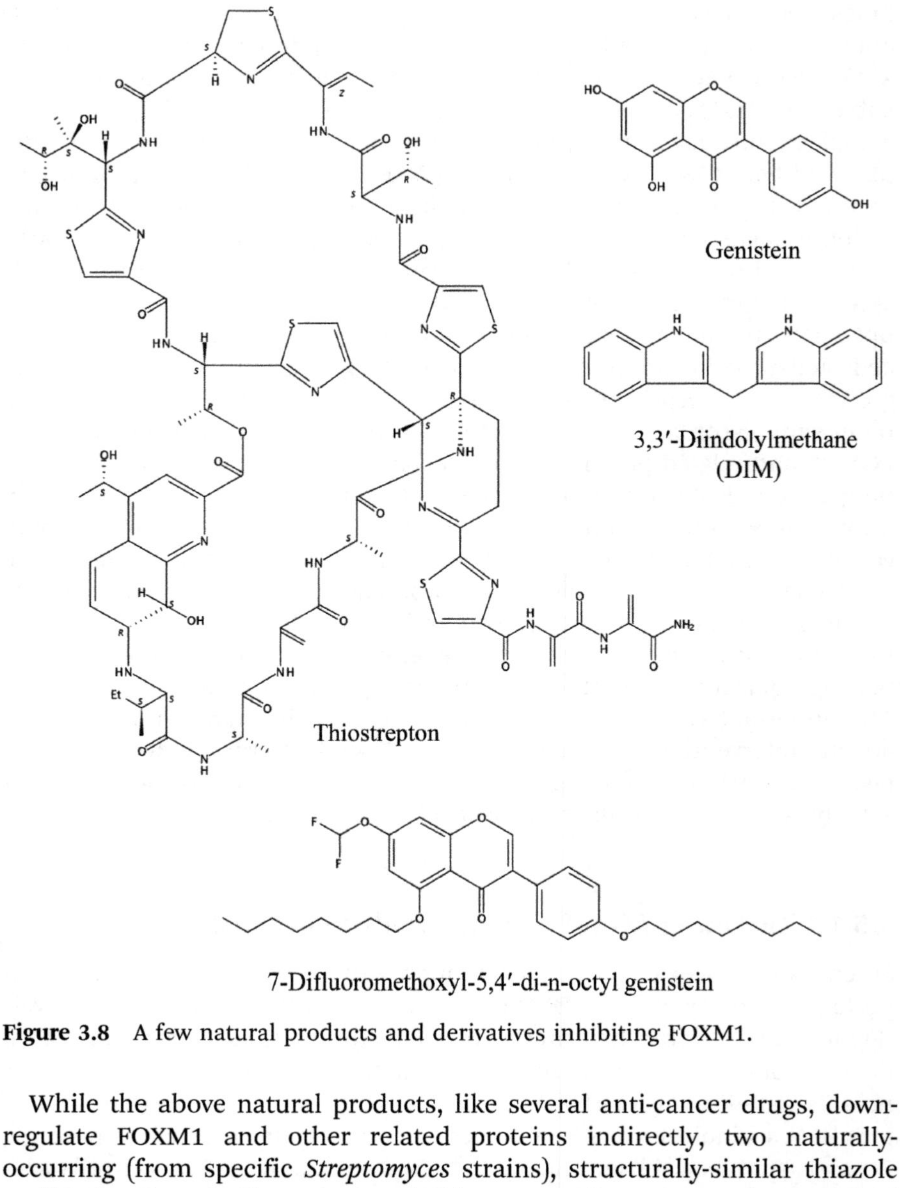

Genistein

3,3′-Diindolylmethane (DIM)

Thiostrepton

7-Difluoromethoxyl-5,4′-di-n-octyl genistein

Figure 3.8 A few natural products and derivatives inhibiting FOXM1.

While the above natural products, like several anti-cancer drugs, down-regulate FOXM1 and other related proteins indirectly, two naturally-occurring (from specific *Streptomyces* strains), structurally-similar thiazole antibiotics, namely Siomycin A and Thiostrepton, have captured a great deal of attention not only because of their potent antibiotic activity but also their exhibiting promising antiproliferative activity against human cancer cells through inhibition of FOXM1.[111] Both of these antibiotics were discovered by the Gartel lab.[118,119] However, several mechanisms have been proposed to account for their activity. For example, studies from the Lam group[120] and the Gartel group[118,119] have shown that siomycin A and thiostrepton inhibit the transcriptional activity and the expression of FOXM, and that they also act as proteasome inhibitors. The Balasubramanian group provided an alternative mode of action for Thiostrepton.[121] They showed that thiostrepton directly binds FoxM1 and thereby prevents it from binding to the promoters of its target genes. This is obviously very encouraging, though further studies are needed to elucidate the precise mechanism of action of these compounds *en route* to developing more selective and specific FoxM1 inhibitors for cancer treatment.

3.6 Conclusion

It is evident from the above examples that a diverse array of naturally occurring chemical scaffolds manifest prominent anti-tumour activity through downregulation of oncogenic transcription factors, including NF-κB, STAT3, HIF-1 and FOXM1. Some of these natural products seem to have privileged structures, therefore showing pleiotropy in action, which makes their desired specificity questionable. However, there is always room for future lead modification that would bias pharmacodynamics towards inhibiting a specific transcription factor. In the majority of the cases, the precise mechanism of action is also yet to be elucidated or validated experimentally. For example, often *in silico* modelling has been used to suggest a specific target of an active phytochemical but the claim is seldom validated through suitable wet experiments (*e.g.* isothermal titration calorimetry with purified transcription factor, site directed mutagenesis *etc.*). Given the historical contribution of natural products in modern drug development, it is encouraging to envisage the possibility of developing potent and efficacious new-generation anti-cancer drugs through iterative lead modifications and extensive SAR profiling together with suitable bioassays and *in vivo* studies.

Acknowledgements

We would like to express our sincere gratitude to the Royal Society, UK and the IDB-Cambridge scholarship.

References

1. J. E. Darnell Jr., *Nat. Rev. Cancer*, 2002, **2**, 740–749.
2. C. Mees, J. Nemunaitis and N. Senzer, *Cancer Gene Ther.*, 2009, **16**, 103–112.
3. D. Hanahan and R. A. Weinberg, *Cell*, 2000, **100**, 57–70.
4. D. Hanahan and R. A. Weinberg, *Cell*, 2011, **144**, 646–674.
5. S. J. Johnston and J. S. Carroll, *Biochim. Biophys. Acta*, 2015, **1855**, 183–192.
6. J. P. Overington, B. Al-Lazikani and A. L. Hopkins, *Nat. Rev. Drug Discovery*, 2006, **5**, 993–996.
7. D. J. Newman, *J. Med. Chem.*, 2008, **51**, 2589–2599.
8. S. C. Gupta, C. Sundaram, S. Reuter and B. B. Aggarwal, *Biochim. Biophys. Acta*, 2010, **1799**, 775–787.

9. Y. Xia, S. Shen and I. M. Verma, *Cancer Immunol. Res.*, 2014, **2**, 823–830.
10. T. D. Gilmore and M. Herscovitch, *Oncogene*, 2006, **25**, 6887–6899.
11. F. Folmer, M. Jaspars, G. Solano, S. Cristofanon, E. Henry, J. Tabudravu, K. Black, D. H. Green, F. C. Kupper, W. Aalbersberg, K. Feussner, M. Dicato and M. Diederich, *Biochem. Pharmacol.*, 2009, **78**, 592–606.
12. S. Luqman and J. M. Pezzuto, *Phytother. Res.*, 2010, **24**, 949–963.
13. M. Karin, Y. Yamamoto and Q. M. Wang, *Nat. Rev. Drug Discovery*, 2004, **3**, 17–26.
14. V. Pande, S. F. Sousa and M. J. Ramos, *Curr. Med. Chem.*, 2009, **16**, 4261–4273.
15. P. Rungeler, V. Castro, G. Mora, N. Goren, W. Vichnewski, H. L. Pahl, I. Merfort and T. J. Schmidt, *Bioorg. Med. Chem.*, 1999, **7**, 2343–2352.
16. A. J. Garcia-Pineres, V. Castro, G. Mora, T. J. Schmidt, E. Strunck, H. L. Pahl and I. Merfort, *J. Biol. Chem.*, 2001, **276**, 39713–39720.
17. B. Siedle, A. J. Garcia-Pineres, R. Murillo, J. Schulte-Monting, V. Castro, P. Rungeler, C. A. Klaas, F. B. Da Costa, W. Kisiel and I. Merfort, *J. Med. Chem.*, 2004, **47**, 6042–6054.
18. Y. F. Xia, B. Q. Ye, Y. D. Li, J. G. Wang, X. J. He, X. Lin, X. Yao, D. Ma, A. Slungaard, R. P. Hebbel, N. S. Key and J. G. Geng, *J. Immunol.*, 2004, **173**, 4207–4217.
19. V. S. Nguyen, X. Y. Loh, H. Wijaya, J. Wang, Q. Lin, Y. Lam, W. S. Wong and Y. K. Mok, *J. Nat. Prod.*, 2015, **78**, 208–217.
20. V. Pande, R. K. Sharma, J. Inoue, M. Otsuka and M. J. Ramos, *J. Comput.-Aided Mol. Des.*, 2003, **17**, 825–836.
21. G. G. Mackenzie, J. M. Delfino, C. L. Keen, C. G. Fraga and P. I. Oteiza, *Biochem. Pharmacol.*, 2009, **78**, 1252–1262.
22. A. C. Castro, L. C. Dang, F. Soucy, L. Grenier, H. Mazdiyasni, M. Hottelet, L. Parent, C. Pien, V. Palombella and J. Adams, *Bioorg. Med. Chem. Lett.*, 2003, **13**, 2419–2422.
23. T. Hideshima, D. Chauhan, P. Richardson, C. Mitsiades, N. Mitsiades, T. Hayashi, N. Munshi, L. Dang, A. Castro, V. Palombella, J. Adams and K. C. Anderson, *J. Biol. Chem.*, 2002, **277**, 16639–16647.
24. D. Wen, Y. Nong, J. G. Morgan, P. Gangurde, A. Bielecki, J. Dasilva, M. Keaveney, H. Cheng, C. Fraser, L. Schopf, M. Hepperle, G. Harriman, B. D. Jaffee, T. D. Ocain and Y. Xu, *J. Pharmacol. Exp. Ther.*, 2006, **317**, 989–1001.
25. S. P. Hehner, T. G. Hofmann, W. Droge and M. L. Schmitz, *J. Immunol.*, 1999, **163**, 5617–5623.
26. B. H. Kwok, B. Koh, M. I. Ndubuisi, M. Elofsson and C. M. Crews, *Chem. Biol.*, 2001, **8**, 759–766.
27. M. C. Liang, S. Bardhan, E. A. Pace, D. Rosman, J. A. Beutler, J. A. Porco Jr. and T. D. Gilmore, *Biochem. Pharmacol.*, 2006, **71**, 634–645.
28. M. C. Liang, S. Bardhan, C. Li, E. A. Pace, J. A. Porco Jr. and T. D. Gilmore, *Mol. Pharmacol.*, 2003, **64**, 123–131.
29. F. Yang, H. S. Oz, S. Barve, W. J. de Villiers, C. J. McClain and G. W. Varilek, *Mol. Pharmacol.*, 2001, **60**, 528–533.

30. H. Zhou, D. M. Monack, N. Kayagaki, I. Wertz, J. Yin, B. Wolf and V. M. Dixit, *J. Exp. Med.*, 2005, **202**, 1327–1332.
31. A. Kumar, H. Wu, L. S. Collier-Hyams, Y. M. Kwon, J. M. Hanson and A. S. Neish, *J. Immunol.*, 2009, **182**, 538–546.
32. M. Hayakawa, H. Miyashita, I. Sakamoto, M. Kitagawa, H. Tanaka, H. Yasuda, M. Karin and K. Kikugawa, *EMBO J.*, 2003, **22**, 3356–3366.
33. R. Sekizawa, S. Ikeno, H. Nakamura, H. Naganawa, S. Matsui, H. Iinuma and T. Takeuchi, *J. Nat. Prod.*, 2002, **65**, 1491–1493.
34. S. Tsukamoto, H. Hirota, M. Imachi, M. Fujimuro, H. Onuki, T. Ohta and H. Yokosawa, *Bioorg. Med. Chem. Lett.*, 2005, **15**, 191–194.
35. K. S. Ahn, G. Sethi, T. H. Chao, S. T. Neuteboom, M. M. Chaturvedi, M. A. Palladino, A. Younes and B. B. Aggarwal, *Blood*, 2007, **110**, 2286–2295.
36. G. Fenteany and S. L. Schreiber, *J. Biol. Chem.*, 1998, **273**, 8545–8548.
37. J. Delic, P. Masdehors, S. Omura, J. M. Cosset, J. Dumont, J. L. Binet and H. Magdelenat, *Br. J. Cancer*, 1998, **77**, 1103–1107.
38. L. Bonfili, V. Cecarini, M. Amici, M. Cuccioloni, M. Angeletti, J. N. Keller and A. M. Eleuteri, *FEBS J.*, 2008, **275**, 5512–5526.
39. D. M. Smith, K. G. Daniel, Z. Wang, W. C. Guida, T. H. Chan and Q. P. Dou, *Proteins*, 2004, **54**, 58–70.
40. D. Chen, M. S. Chen, Q. C. Cui, H. Yang and Q. P. Dou, *Front. Biosci.*, 2007, **12**, 1935–1945.
41. N. Lu, H. Hui, H. Yang, K. Zhao, Y. Chen, Q. D. You and Q. L. Guo, *Eur. J. Pharm. Sci.*, 2013, **49**, 220–226.
42. X. Li, S. Liu, H. Huang, N. Liu, C. Zhao, S. Liao, C. Yang, Y. Liu, C. Zhao, S. Li, X. Lu, C. Liu, L. Guan, K. Zhao, X. Shi, W. Song, P. Zhou, X. Dong, H. Guo, G. Wen, C. Zhang, L. Jiang, N. Ma, B. Li, S. Wang, H. Tan, X. Wang, Q. P. Dou and J. Liu, *Cell Rep.*, 2013, **3**, 211–222.
43. V. Khedgikar, P. Kushwaha, J. Gautam, A. Verma, B. Changkija, A. Kumar, S. Sharma, G. K. Nagar, D. Singh, P. K. Trivedi, N. S. Sangwan, P. R. Mishra and R. Trivedi, *Cell Death Dis.*, 2013, **4**, e778.
44. H. Jiang, J. Sun, Q. Xu, Y. Liu, J. Wei, C. Y. Young, H. Yuan and H. Lou, *Cell Death Dis.*, 2013, **4**, e761.
45. H. Yu and R. Jove, *Nat. Rev. Cancer*, 2004, **4**, 97–105.
46. D. Germain and D. A. Frank, *Clin. Cancer Res.*, 2007, **13**, 5665–5669.
47. G. Miklossy, T. S. Hilliard and J. Turkson, *Nat. Rev. Drug Discovery*, 2013, **12**, 611–629.
48. N. Arumuggam, N. A. Bhowmick and H. P. Rupasinghe, *Phytother. Res.*, 2015, **29**, 805–817.
49. J. Sun, M. A. Blaskovich, R. Jove, S. K. Livingston, D. Coppola and S. M. Sebti, *Oncogene*, 2005, **24**, 3236–3245.
50. N. H. Thoennissen, G. B. Iwanski, N. B. Doan, R. Okamoto, P. Lin, S. Abbassi, J. H. Song, D. Yin, M. Toh, W. D. Xie, J. W. Said and H. P. Koeffler, *Cancer Res.*, 2009, **69**, 5876–5884.
51. M. A. Blaskovich, J. Sun, A. Cantor, J. Turkson, R. Jove and S. M. Sebti, *Cancer Res.*, 2003, **63**, 1270–1279.

52. Y. Dong, B. Lu, X. Zhang, J. Zhang, L. Lai, D. Li, Y. Wu, Y. Song, J. Luo, X. Pang, Z. Yi and M. Liu, *Carcinogenesis*, 2010, **31**, 2097–2104.
53. D. S. Shin, H. N. Kim, K. D. Shin, Y. J. Yoon, S. J. Kim, D. C. Han and B. M. Kwon, *Cancer Res.*, 2009, **69**, 193–202.
54. Y. J. Jeon, S. N. Jung, J. Yun, C. W. Lee, J. Choi, Y. J. Lee, D. C. Han and B. M. Kwon, *Cancer Sci.*, 2015, **106**, 413–420.
55. K. S. Ahn, G. Sethi, B. Sung, A. Goel, R. Ralhan and B. B. Aggarwal, *Cancer Res.*, 2008, **68**, 4406–4415.
56. X. Chen, Y. Du, J. Nan, X. Zhang, X. Qin, Y. Wang, J. Hou, Q. Wang and J. Yang, *PLoS One*, 2013, **8**, e63697.
57. J. Zhang, N. Zhu, Y. Du, Q. Bai, X. Chen, J. Nan, X. Qin, X. Zhang, J. Hou, Q. Wang and J. Yang, *Mol. Pharmacol.*, 2015, **87**, 572–581.
58. Y. Wang, X. Ma, S. Yan, S. Shen, H. Zhu, Y. Gu, H. Wang, G. Qin and Q. Yu, *Cancer Res.*, 2009, **69**, 7302–7310.
59. C. Agarwal, A. Tyagi, M. Kaur and R. Agarwal, *Carcinogenesis*, 2007, **28**, 1463–1470.
60. J. E. Jung, H. S. Kim, C. S. Lee, D. H. Park, Y. N. Kim, M. J. Lee, J. W. Lee, J. W. Park, M. S. Kim, S. K. Ye and M. H. Chung, *Carcinogenesis*, 2007, **28**, 1780–1787.
61. A. Kotha, M. Sekharam, L. Cilenti, K. Siddiquee, A. Khaled, A. S. Zervos, B. Carter, J. Turkson and R. Jove, *Mol. Cancer Ther.*, 2006, **5**, 621–629.
62. S. Nam, R. Buettner, J. Turkson, D. Kim, J. Q. Cheng, S. Muehlbeyer, F. Hippe, S. Vatter, K. H. Merz, G. Eisenbrand and R. Jove, *Proc. Natl. Acad. Sci. U. S. A.*, 2005, **102**, 5998–6003.
63. H. Y. Kim, E. J. Park, E. H. Joe and I. Jou, *J. Immunol.*, 2003, **171**, 6072–6079.
64. S. R. Walker, M. Chaudhury, E. A. Nelson and D. A. Frank, *Mol. Pharmacol.*, 2010, **78**, 903–908.
65. A. Kumar and U. Bora, *Bioinformation*, 2012, **8**, 988–993.
66. G. Sethi, S. Chatterjee, P. Rajendran, F. Li, M. K. Shanmugam, K. F. Wong, A. P. Kumar, P. Senapati, A. K. Behera, K. M. Hui, J. Basha, N. Natesh, J. M. Luk and T. K. Kundu, *Mol. Cancer*, 2014, **13**, 66.
67. J. K. Kim, J. Y. Kim, H. J. Kim, K. G. Park, R. A. Harris, W. J. Cho, J. T. Lee and I. K. Lee, *PLoS One*, 2013, **8**, e80391.
68. L. P. Yco, G. Mocz, J. Opoku-Ansah and A. S. Bachmann, *Biochem. Insights*, 2014, **7**, 1–13.
69. N. Wu, J. Liu, X. Zhao, Z. Yan, B. Jiang, L. Wang, S. Cao, D. Shi and X. Lin, *Tumour Biol.*, 2015, **36**, 9667–9676.
70. J. Mantaj, S. M. Rahman, B. Bokshi, C. M. Hasan, P. J. Jackson, R. B. Parsons and K. M. Rahman, *Org. Biomol. Chem.*, 2015, **13**, 3882–3886.
71. M. Weidler, J. Rether, T. Anke and G. Erkel, *FEBS Lett.*, 2000, **484**, 1–6.
72. Y. K. Lee, C. R. Isham, S. H. Kaufman and K. C. Bible, *Mol. Cancer Ther.*, 2006, **5**, 138–148.
73. G. L. Semenza, *Curr. Pharm. Des.*, 2009, **15**, 3839–3843.
74. M. H. Rabinowitz, *J. Med. Chem.*, 2013, **56**, 9369–9402.
75. J. M. Brown and A. J. Giaccia, *Cancer Res.*, 1998, **58**, 1408–1416.

76. M. Nepal, H. J. Choi, B. Y. Choi, S. L. Kim, J. H. Ryu, H. Kim do, Y. H. Lee and Y. Soh, *Eur. J. Pharmacol.*, 2012, **691**, 28–37.
77. H. J. Choi, J. S. Eun, D. K. Kim, R. H. Li, T. Y. Shin, H. Park, N. P. Cho and Y. Soh, *Eur. J. Pharmacol.*, 2008, **579**, 58–65.
78. D. B. Jung, H. J. Lee, S. J. Jeong, H. J. Lee, E. O. Lee, Y. C. Kim, K. S. Ahn, C. Y. Chen and S. H. Kim, *Biol. Pharm. Bull.*, 2011, **34**, 850–855.
79. H. O. Jun, Y. Kim, Y. W. Kwon, S. S. Hong, K. W. Kim, J. Shin and T. Y. Kim, *FEBS Lett.*, 2007, **581**, 4977–4982.
80. J. Zhou, S. Zhang, J. Xue, J. Avery, J. Wu, S. E. Lind and W. Q. Ding, *J. Biol. Chem.*, 2012, **287**, 35161–35169.
81. J. S. Isaacs, Y. J. Jung, E. G. Mimnaugh, A. Martinez, F. Cuttitta and L. M. Neckers, *J. Biol. Chem.*, 2002, **277**, 29936–29944.
82. J. S. Isaacs, Y. J. Jung and L. Neckers, *J. Biol. Chem.*, 2004, **279**, 16128–16135.
83. D. M. Katschinski, L. Le, S. G. Schindler, T. Thomas, A. K. Voss and R. H. Wenger, *Cell. Physiol. Biochem.*, 2004, **14**, 351–360.
84. E. Minet, D. Mottet, G. Michel, I. Roland, M. Raes, J. Remacle and C. Michiels, *FEBS Lett.*, 1999, **460**, 251–256.
85. A. Khandelwal, V. M. Crowley and B. S. Blagg, *Med. Res. Rev.*, 2016, **36**, 92–118.
86. N. J. Mabjeesh, D. E. Post, M. T. Willard, B. Kaur, E. G. Van Meir, J. W. Simons and H. Zhong, *Cancer Res.*, 2002, **62**, 2478–2482.
87. X. Song, J. Yao, F. Wang, M. Zhou, Y. Zhou, H. Wang, L. Wei, L. Zhao, Z. Li, N. Lu and Q. Guo, *Toxicol. Appl. Pharmacol.*, 2013, **271**, 144–155.
88. M. Osada, S. Imaoka and Y. Funae, *FEBS Lett.*, 2004, **575**, 59–63.
89. B. Fu, J. Xue, Z. Li, X. Shi, B. H. Jiang and J. Fang, *Mol. Cancer Ther.*, 2007, **6**, 220–226.
90. S. H. Lee, J. G. Jee, J. S. Bae, K. H. Liu and Y. M. Lee, *J. Cell. Physiol.*, 2015, **230**, 853–862.
91. E. Hur, H. H. Kim, S. M. Choi, J. H. Kim, S. Yim, H. J. Kwon, Y. Choi, D. K. Kim, M. O. Lee and H. Park, *Mol. Pharmacol.*, 2002, **62**, 975–982.
92. B. Yu, Z. H. Miao, Y. Jiang, M. H. Li, N. Yang, T. Li and J. Ding, *Cancer Res.*, 2009, **69**, 7704–7712.
93. B. Yu, M. H. Li, W. Wang, Y. Q. Wang, Y. Jiang, S. P. Yang, J. M. Yue, J. Ding and Z. H. Miao, *J. Mol. Med. (Berl)*, 2012, **90**, 971–981.
94. H. H. Lin, C. W. Tsai, F. P. Chou, C. J. Wang, S. W. Hsuan, C. K. Wang and J. H. Chen, *Toxicol. Appl. Pharmacol.*, 2011, **250**, 336–345.
95. M. C. Chen, C. F. Lee, W. H. Huang and T. C. Chou, *Biochem. Pharmacol.*, 2013, **85**, 1278–1287.
96. S. H. Kim, J. W. Jeong, J. A. Park, J. W. Lee, J. H. Seo, B. K. Jung, M. K. Bae and K. W. Kim, *Oncol. Rep.*, 2007, **17**, 647–651.
97. D. Kong, E. J. Park, A. G. Stephen, M. Calvani, J. H. Cardellina, A. Monks, R. J. Fisher, R. H. Shoemaker and G. Melillo, *Cancer Res.*, 2005, **65**, 9047–9055.
98. J. V. Formica and M. J. Waring, *Antimicrob. Agents Chemother.*, 1983, **24**, 735–741.

99. A. L. Kung, S. D. Zabludoff, D. S. France, S. J. Freedman, E. A. Tanner, A. Vieira, S. Cornell-Kennon, J. Lee, B. Wang, J. Wang, K. Memmert, H. U. Naegeli, F. Petersen, M. J. Eck, K. W. Bair, A. W. Wood and D. M. Livingston, *Cancer Cell*, 2004, **6**, 33–43.
100. K. M. Cook, S. T. Hilton, J. Mecinovic, W. B. Motherwell, W. D. Figg and C. J. Schofield, *J. Biol. Chem.*, 2009, **284**, 26831–26838.
101. M. K. Jayatunga, S. Thompson, T. C. McKee, M. C. Chan, K. M. Reece, A. P. Hardy, R. Sekirnik, P. T. Seden, K. M. Cook, J. B. McMahon, W. D. Figg, C. J. Schofield and A. D. Hamilton, *Eur. J. Med. Chem.*, 2015, **94**, 509–516.
102. S. T. Chan, P. R. Patel, T. R. Ransom, C. J. Henrich, T. C. McKee, A. K. Goey, K. M. Cook, W. D. Figg, J. B. McMahon, M. J. Schnermann and K. R. Gustafson, *J. Am. Chem. Soc.*, 2015, **137**, 5569–5575.
103. D. Wu, R. Zhang, R. Zhao, G. Chen, Y. Cai and J. Jin, *PLoS One*, 2013, **8**, e62014.
104. Y. J. Choi, H. Choi, C. H. Cho and J. W. Park, *J. Nat. Med.*, 2011, **65**, 344–352.
105. K. Lee, H. Zhang, D. Z. Qian, S. Rey, J. O. Liu and G. L. Semenza, *Proc. Natl. Acad. Sci. U. S. A.*, 2009, **106**, 17910–17915.
106. T. Klimova and N. S. Chandel, *Cell Death Differ.*, 2008, **15**, 660–666.
107. D. G. Nagle and Y. D. Zhou, *Phytochem. Rev.*, 2009, **8**, 415–429.
108. W. Korver, J. Roose, K. Heinen, D. O. Weghuis, D. de Bruijn, A. G. van Kessel and H. Clevers, *Genomics*, 1997, **46**, 435–442.
109. J. Laoukili, M. Stahl and R. H. Medema, *Biochim. Biophys. Acta*, 2007, **1775**, 92–102.
110. I. C. Wang, Y. J. Chen, D. Hughes, V. Petrovic, M. L. Major, H. J. Park, Y. Tan, T. Ackerson and R. H. Costa, *Mol. Cell. Biol.*, 2005, **25**, 10875–10894.
111. M. Halasi and A. L. Gartel, *Biochem. Pharmacol.*, 2013, **85**, 644–652.
112. Y. S. Kim and J. A. Milner, *J. Nutr. Biochem.*, 2005, **16**, 65–73.
113. A. Ahmad, S. Ali, Z. Wang, A. S. Ali, S. Sethi, W. A. Sakr, A. Raz and K. M. Rahman, *Int. J. Cancer*, 2011, **129**, 1781–1791.
114. Y. X. Ning, Q. X. Li, K. Q. Ren, M. F. Quan and J. G. Cao, *Oncol. Lett.*, 2014, **8**, 295–300.
115. T. Tian, J. Li, B. Li, Y. Wang, M. Li, D. Ma and X. Wang, *Tumour Biol.*, 2014, **35**, 4137–4145.
116. Z. Wang, A. Ahmad, S. Banerjee, A. Azmi, D. Kong, Y. Li and F. H. Sarkar, *Pharm. Res.*, 2010, **27**, 1159–1168.
117. H. L. Xiang, F. Liu, M. F. Quan, J. G. Cao and Y. Lv, *World J. Gastroenterol.*, 2012, **18**, 4618–4626.
118. U. G. Bhat, M. Halasi and A. L. Gartel, *PLoS One*, 2009, **4**, e5592.
119. S. K. Radhakrishnan, U. G. Bhat, D. E. Hughes, I. C. Wang, R. H. Costa and A. L. Gartel, *Cancer Res.*, 2006, **66**, 9731–9735.
120. J. M. Kwok, S. S. Myatt, C. M. Marson, R. C. Coombes, D. Constantinidou and E. W. Lam, *Mol. Cancer Ther.*, 2008, **7**, 2022–2032.
121. N. S. Hegde, D. A. Sanders, R. Rodriguez and S. Balasubramanian, *Nat. Chem.*, 2011, **3**, 725–731.

CHAPTER 4

Pyrrolobenzodiazepines as Transcription Factor Inhibitors: An Overview

DAVID B. CORCORAN, DAVID E. THURSTON AND KHONDAKER MIRAZ RAHMAN*

School of Cancer and Pharmaceutical Science, King's College London, SE1 9NH, and Femtogenix Limited, Biopark, Broadwater Road, Welwyn Garden City AL7 3AX, UK
*Email: k.miraz.rahman@kcl.ac.uk

4.1 Transcription Factors and Their Role In Gene Expression

Before a detailed examination of pyrrolobenzodiazepine (PBD) structures that can serve as transcription factors is undertaken, it is important to understand the role of transcription factors in the control of gene expression and how they may contribute to the pathogenesis of malignancy. Transcription factors are large polypeptide molecules which bind to short (6–12 bp) DNA sequences known as enhancer regions,[1] facilitating or repressing transcription of genes by RNA polymerase II. As such powerful influencers of cellular genomic activity, the activity of transcription factors must be highly controlled though a number of different methods.

The cellular concentration and activity of transcription factors themselves is tightly regulated by the cell.[2] Some transcription factors are solely expressed in certain tissues or at key points in cellular development.[3] The

Drug Discovery Series No. 65
Small-molecule Transcription Factor Inhibitors in Oncology
Edited by Khondaker Miraz Rahman and David E. Thurston

Published by the Royal Society of Chemistry, www.rsc.org

highly conserved forkhead box a (FOXA) transcription factors provide a good example of transcriptional control through both time- and tissue-specific expression. FOXA transcription factors play an essential role in transcriptional regulation during the early development phase of several mammalian organs. Upon maturity of an organism, they are selectively expressed in the liver, pancreas and lung tissues to maintain cellular homeostasis.[4,5] Other transcription factors exist constitutively in cells in an inactive form. Upon activation *via* intracellular or extracellular signalling or by undergoing some form of modification, such as phosphorylation, they then act to influence gene expression. The Janus kinase–signal transducer and activator of transcription (JAK–STAT) transcription factor activation pathway is a good example. The STAT transcription factors are important mediators of cellular growth and development processes as well as the control of apoptosis.[6–8] In their inactive state, they exist as monomers, incapable of nuclear localisation and binding. The binding of exogenous messengers, such as interferons, to transmembrane receptors causes the formation of multi-receptor complexes. The formation of these complexes activates associated kinases (JAKs). The JAK proteins then phosphorylate the transmembrane receptors, attracting cytosolic STAT monomers, which are, in turn, phosphorylated. The phosphorylation of STAT causes its dimerization, allowing for translocation to the nucleus and control of cellular transcription through binding to promoter or enhancer elements. This activation pathway is summarised in Figure 4.1.

The complex control of transcription factor synthesis and activation described above is further complemented by concurrent regulation of their binding to DNA. In many cases, transcription factors bind in multiple

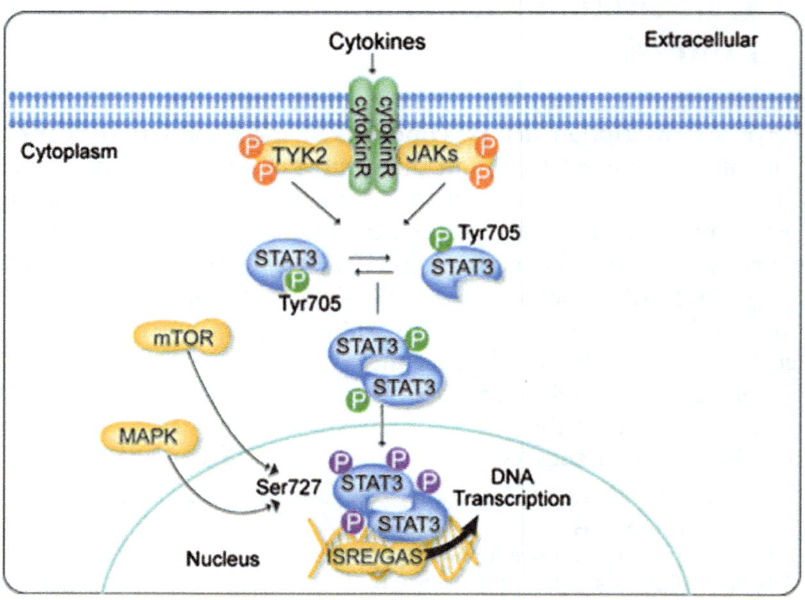

Figure 4.1 JAK–STAT signalling pathway.

combinations to the enhancer regions of groups of related genes, providing the ability to subtly control and tailor phenotypic outcomes to specific cellular requirements. For example, DNA binding of the transcription factor Lhx1 [lineage 11–insulin enhancer protein–mechanosensory protein 3 (LIM) domain homeobox 1] during the early stages of mammalian development demonstrates the effect of this interplay between transcription factors. Lhx1 has been shown to bind to enhancer regions also bound by the Foxa2 and Otx2 (orthodenticle homeobox 2) transcription factors, indicating a combined role for these transcription factors in early tissue differentiation.[9] The effects of transcription factor binding may also be influenced by the architecture of the enhancer region itself. Alterations to the DNA sequence of an enhancer or differences in the spatial DNA arrangement of the enhancer structure represent a further cellular control point of transcription factor activity, a concept reviewed by Spitz and Furlong.[1]

Although transcription factors differ significantly in their structures, functions and methods of activation, certain motifs are common to all transcription factors.[2,3,10] For example, all possess a DNA-binding motif consisting of a peptide sequence capable of recognizing and binding to specific enhancer regions of DNA across the genome. Another region commonly found in many transcription factors is a trans-activating domain. Following binding of the factor to target areas of the DNA duplex, the trans-activating domain of the factor interacts with the basal transcription complex, a group of proteins including RNA polymerase II required for transcription of the gene. This interaction may promote or repress the stability of the basal complex, thus modifying the rate of transcription.

4.2 Transcription Factors as Mediators of Carcinogenesis

Transcription factors provide the machinery through which a cell controls the expression of its genome in a regulated manner. However, the fundamental vulnerability of DNA to mutagenic changes presents a challenge to this control. In many respects, the complexity of the transcription factor activation processes acts as a double-edged sword. The presence of multi-protein transcription factor activation cascades and the large array of multifunctional transcription factors undoubtedly add to the precision of genetic expression. However, this also increases the chances of dysfunction, as mutations to any one of the genes involved in such processes may affect transcription. The dysregulation of transcription factor activity has been linked to the development of numerous types of cancer.[11]

The effect of mutations to the RAS (rat sarcoma) transcription factor activation pathway (which occur in up to 30% of human cancers) provides a good insight into how transcription factors may become potential mediators of carcinogenesis.[12,13] The *Ras* gene family expresses proteins known as RAS kinases. These protein activators act as critical mediators between membrane receptors that receive growth signals such as the endothelial growth factor receptor (EGFR), and their ultimate downstream effectors [*e.g.* transcription factors such as activator protein 1 (AP-1)]. Gene expression mediated by AP-1 is essential for cell cycle progression. In a normal cell, RAS proteins exist in both an active and inactive state, thus controlling the transcription activator cascade effectively. However, mutations to *NRAS* (neuroblastoma RAS), *HRAS* (Harvey RAS) and *KRAS2* (Kirsten RAS 2) genes may produce mutated RAS, which remains in its activated (guanosine triphosphate-bound) form. These modified RAS proteins cannot regulate the activity of AP-1, which remains in an activated state. Mutations of other effectors of the RAS transcription factor activation pathway have also been implicated in carcinogenesis and have been reviewed by Downward.[12]

The dysregulation of the cytosolic transcription factor nuclear factor kappa B (NF-κB) activation process further illustrates the potential for transcription factors to underpin cellular transformation. Comprised of a dimer from any two of five different subunits, NF-κB resides in the cytosol of the cell bound to an inhibitory protein (IkB) which prevents nuclear localisation. Upon stimulation by a member of a wide array of exogenous and endogenous signalling molecules including tumour necrosis factor alpha, a protein phosphorylation cascade involving a variety of different activation pathways culminates in the liberation of NF-κB from its inhibitory complex through a protein known as IkB kinase (IKK). This allows translocation to the nucleus followed by binding to DNA enhancer regions. The activation pathway of NF-κB has been reviewed by Ghosh and colleagues,[14,15] and the signalling pathway is detailed in Figure 4.2. NF-κB is central to the

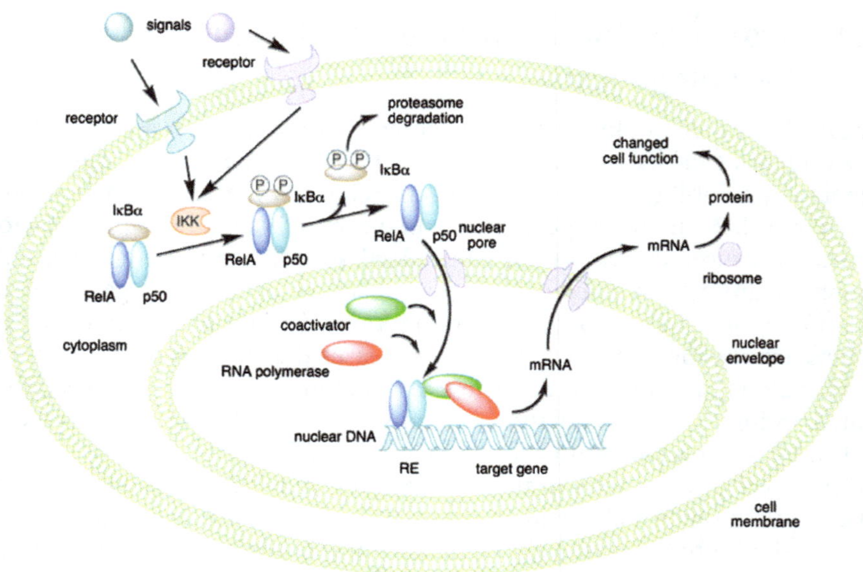

Figure 4.2 NF-κB signalling pathway.

transcription of genes that control inflammatory immune response and cellular development. Most importantly from the perspective of its role in cancer, NF-κB also controls the expression of anti-apoptotic factors such as IAP (inhibitor of apoptosis proteins), Bcl-2 (B-cell lymphoma 2), Bcl-X_L (B-cell leukaemia x large) A20 and NR-1 (*N*-methyl-D-aspartate receptor 1) proteins.[16,17] Overactive NF-κB has been found to play a role in the pathogenesis of numerous cancers, especially blood-borne malignancies. Acute lymphoblastic leukaemia (ALL), chronic lymphocytic leukaemia (CLL), acute and chronic myeloid leukaemia (CML) and multiple myeloma (MM) have all been linked to abnormal NF-κB activity.[18–23]

4.3 Targeting Transcription Factors: Proceeding With Caution

The examples described above illustrate how the modification of transcription factor activity represents a clear target for the treatment of many cancer types. However, there are significant challenges to effectively altering transcription factor functionality. In some cases, the initial dysregulation of one transcription factor may lead to the abnormal activity of others due to the inherent redundancy and cross-talk capacity between different transcription factor pathways. This could potentially lead to unwanted off-target effects for a potential therapeutic agent operating through this mechanism. For example, in some lung cancers, abnormal activity of the RAS activation pathway induces a p53-independent apoptotic pathway.[24,25] However, this apoptotic pathway is superseded by activation of the NF-κB pathway, which prevents the cell entering apoptosis. Inhibition of NF-κB was toxic in some experiments with these cells, while many inhibitors of the RAS pathway itself failed. Essentially, cross-talk between various gene expression pathways may render potential targets difficult to delineate without a deeper understanding of the often complex activation mechanisms involved.

Further issues arise when single abnormal components of gene expression are targeted. The failure of farnesyl transferase inhibitors (FTIs) as modulators of KRAS function provides a good example of the redundancy built into transcription factor activation cascades. RAS proteins require modification with a sesquiterpene farnesyl group in order to associate with the cellular membrane and to function effectively.[12,13,26] Thus, blockade of the farnesyl transferase enzyme was viewed as an exciting chemotherapeutic target in patients suffering from KRAS-mediated malignancy, and several agents entered clinical trials.[27] However, clinical studies showed disappointing efficacy, most probably due to the ability of geranylgeranylated KRAS to also be able to modulate gene expression.[28] Therefore, redundancy within the activation pathway apparently rendered the targeted therapy ineffective. For example, the farnesyltransferase inhibitor tipifarnib (Zarnestra™) developed by Johnson & Johnson was refused marketing approval by the FDA in 2005 (Figure 4.3). Resistance to the inhibition of many

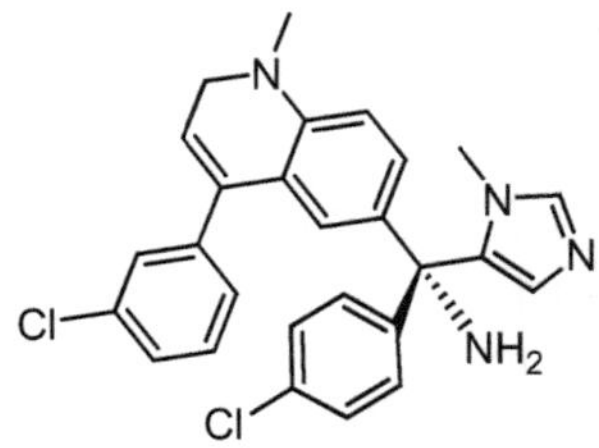

Figure 4.3 The structure of tipifarnib, a farnesyl transferase inhibitor that recently failed phase III clinical trials for use in colon cancer.

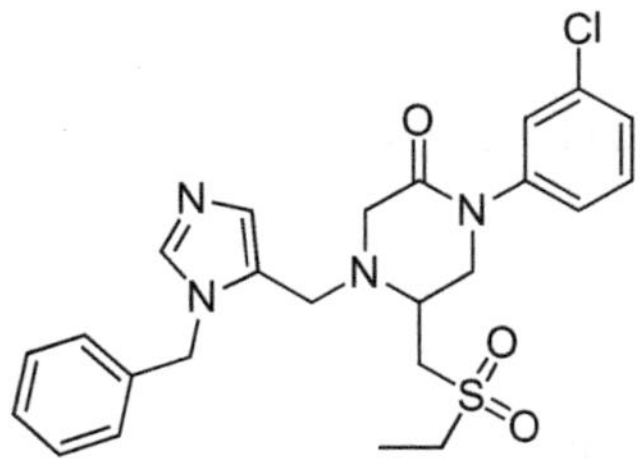

Figure 4.4 Structure of L778123, a failed dual farnesyl/geranylgeranyl transferase inhibitor.

other members of the RAS activation cascade such as RAF (rapidly accelerated fibrosarcoma) and MEK (melanoma kinase) has also recently been shown to be mediated through cellular cross-talk and redundancy.[29]

Finally, it is important to note that although transcription factor inhibition may dramatically affect the growth of tumour cells, many transcription factors are essential for maintaining normal cellular homeostasis. Many effective inhibitors of critical transcription factor signalling pathways have proven to be too toxic for clinical use due to unacceptable side effects. For example, after the issues with farnesyl transferase inhibitors were identified, combination inhibition of farnesyl and geranylgeranyl transferase in RAS-mediated cancers was attempted.[30] However, they proved too toxic for clinical use (Figure 4.4). Inhibitors of the various components of the NF-κB pathway have been developed, including IKKb inhibitors. These inhibitors were associated with a paradoxical increase in immune response to bacterial endotoxin, thus increasing the risk of septic shock.[31]

4.4 PBDs as Potential Transcription Factor Inhibitors

Members of the pyrrolo[2, 1-*c*][1, 4]benzodiazepine (PBD) class of compounds have been extensively explored as potential chemotherapeutic agents due to their ability to bind covalently to specific DNA sequences within the DNA minor groove. PBDs are comprised of a three-ring structure containing benzodiazepine A and B rings fused to a third pyrrolidine C ring. The core structure of a PBD is shown in Figure 4.5, which also illustrates the numbering system used to describe substituted analogues.

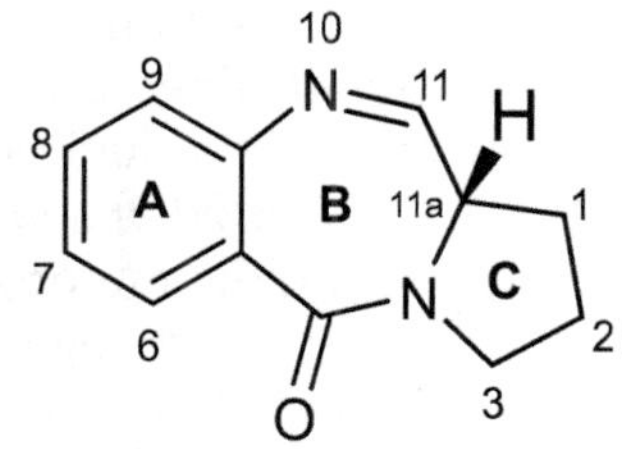

Figure 4.5 The core PBD structure, with rings labelled and key atoms numbered.

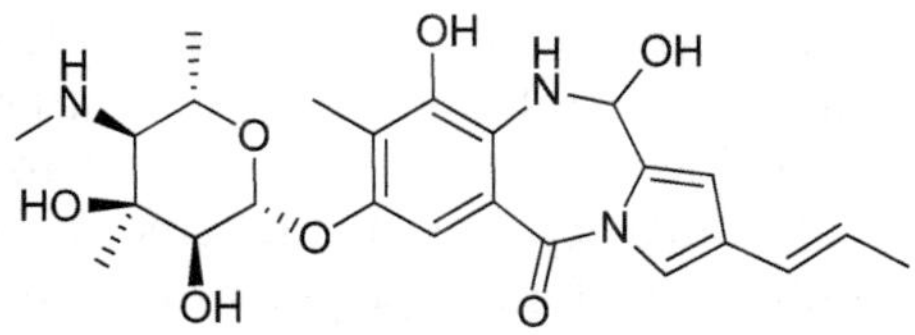

Sibiromycin

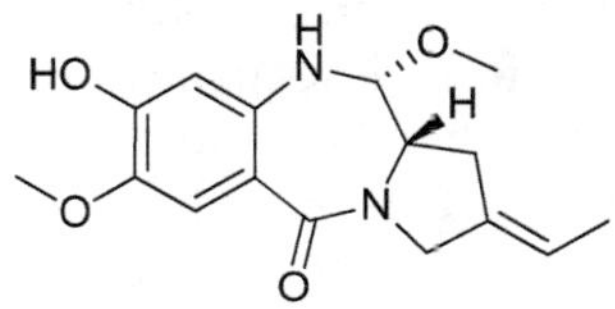

Tomaymycin

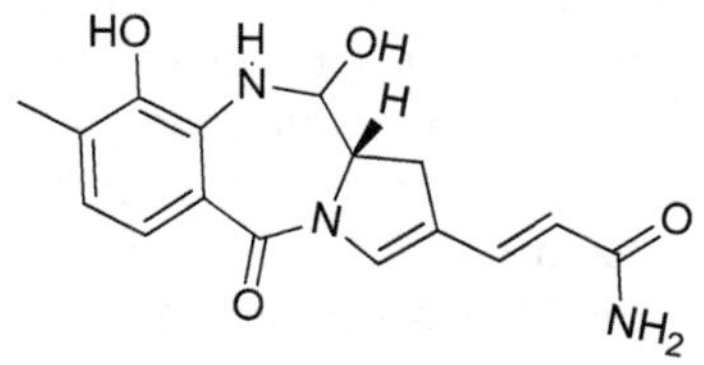

Anthramycin

Figure 4.6 Structures of the naturally-occurring PBDs sibiromycin, tomaymycin and anthramycin.

Naturally-occurring PBD compounds such as anthramycin, tomaymycin and sibiromycin (Figure 4.6) were the first to be studied after their isolation as secondary metabolites from *Streptomyces* bacteria.[32,33] PBD structures also possess antibacterial properties[32,34] and this has led to studies of synthetic derivatives as potential anti-infective therapies.[35] However, in the past their toxicity profiles[36] have limited their clinical development as either anti-infective or anti-cancer agents, although they are currently finding success as cytotoxic payloads for antibody–drug conjugates (ADCs).[117] Current research on the PBDs aims to increase safety while maintaining and enhancing biological potency.

Two core features of the PBD molecule confer the capacity for DNA sequence-specific binding within the DNA minor groove. First, *S*-stereochemistry at the C11a-position confers a right-handed longitudinal twist upon the molecule, which allows a neat fit within the curved minor groove of the DNA double helix. The importance of this stereochemistry has been confirmed through the synthesis of PBD analogues containing C11a *R*-stereochemistry which do not interact with DNA and are biologically inactive.[32,37] While the *S*-stereochemistry at the C11a-position of the PBD core allows for positioning within the DNA minor groove, it is the presence of an interconvertible imine–carbinolamine–carbinolamine methyl ether functionality at the N10–C11-position of the molecule that allows covalent DNA binding. The electrophilic C11 position is attacked by a C2-NH_2 functional group of a guanine residue within the minor groove, thus forming an aminal covalent bond (Figure 4.7).[38,39] Although this covalent bond is thought to be highly stable with the PBD molecules being held in place in the minor groove by hydrogen bonding and hydrophobic interactions *etc.* there is some evidence that the adduct may be reversible, especially at higher temperatures and/or low pH.[40,41]

The mechanism of action of unmodified pyrrolobenzodiazepines, which have limited sequence-selectivity, resembles that of traditional alkylating agents that have not been utilised as modulators of transcription factor activity. For both types of molecules, cellular cytotoxicity results from mechanisms including non-specific inhibition of RNA polymerase[42] and endonucleases[43] and effects on protein synthesis. However, the preferential binding of pyrrolobenzodiazepines to certain guanine-containing DNA motifs presents the possibility of selective inhibition of certain DNA-processing mechanisms. For example, through DNA footprinting studies, it has been demonstrated that from a thermodynamic perspective PBDs prefer to bind to guanine residues flanked by purines (Pu), while guanines flanked by pyrimidines (Py) represent unfavourable binding sites.[44] Interestingly, from a kinetic perspective, Py–G–Py sequences appear to be targeted first but with a gradual transition to the thermodynamically-preferred sequences over a 24-hour period.[45] This sequence-selectivity of DNA interaction raises the possibility that PBD-type molecules may be developed to interfere with transcription factor activity at the DNA level. Inhibition of transcription factor signalling at the end of the signalling process (*i.e.* at the point at which the transcription factor itself interacts with the gene) is an attractive proposition, as it should avoid cross-talk and redundancy issues, which can frustrate the inhibition of activator and signalling proteins upstream of the transcription factor DNA-binding event.

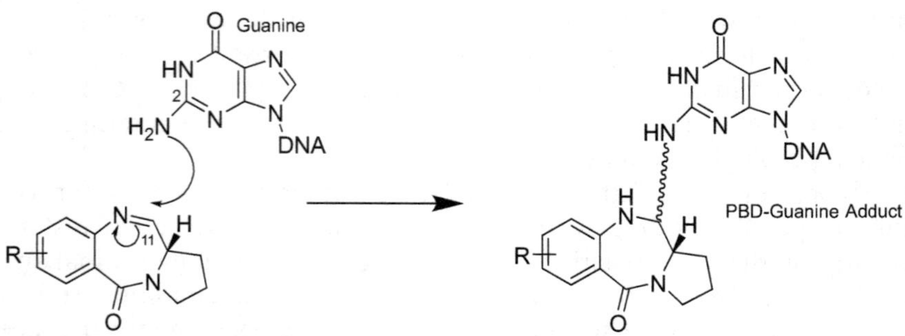

Figure 4.7 Formation of a covalent bond between the C2-NH_2 group of a guanine residue and the electrophilic C11-position of a PBD within the minor groove of DNA.

In order to develop PBD-type molecules as transcription factor inhibitors, the basic PBD core structure requires modification to allow coverage of a greater number of DNA base pairs. The basic PBD core spans three DNA base pairs (*i.e.* the covalently modified guanine flanked by a base pair on either side). As transcription factor enhancer sequences generally span 6–12 base pairs[1] such adducts would be too short to provide effective inhibition of transcription factor binding, and would also lack selectivity, as large numbers of sites across the genome would be affected. Therefore, the goals of the development of PBDs as transcription factor inhibitors are to join the basic core structure to other molecular moieties with DNA sequence-selectivity to cover a greater number of bases pairs, to enhance sequence selectivity (*i.e.* in particular to target specific transcription factor binding sites) and to reduce the overall number of adducts across the genome (*i.e.* theoretically reducing off-target effects and systemic toxicity) (Figure 4.8).

4.5 Distamycin-based PBD Transcription Factor Inhibitors: One Small Step

The *Streptomyces* genus of bacteria has contributed greatly to the development of PBD-containing transcription factor inhibitors. *Streptomyces* bacteria not only provided the first PBD molecules themselves, but also provided distamycin, a member of another family of DNA-interactive molecules, which has been attached to a PBD to extend the DNA base-pair span (*i.e.* to about five base pairs) and enhance sequence recognition properties. The guanidinium-containing polypyrrole structure of distamycin is shown in Figure 4.9.

Unlike the binding of PBD molecules to DNA, which involves covalent binding, distamycin-type compounds interact through hydrogen bonding and van der Waals interactions. These non-covalent, reversible interactions create a preference for A/T-rich sequences.[46,47] Through this interaction with DNA, distamycin-type molecules cause cytotoxic effects thought to be due, in part, to disruption of the DNA processing enzymes topoisomerase I and II[48,49] Distamycin has also been reported to have inhibitory properties toward some transcription factor activity. For example, it has been reported to prevent binding of the transcription factors SRF (serum response factor) and MEF2 (myocyte enhancer factor 2) to their AT-rich binding sequences in muscle cells, thus preventing cellular differentiation.[50] Another study reported a reduction of binding of the OTF-1 (octamer transcription factor 1) and NFE-1 (nuclear factor erythroid 1) transcription factors to AT-rich promoter or enhancer sequences in cells that had been incubated with distamycin.[51]

The DNA sequence-selective and transcription-factor-inhibitory properties of distamycin made it an ideal candidate for conjugation to the covalently interactive PBD core structure. The first such conjugate was synthesised by Baraldi and co-workers[52] and is illustrated in Figure 4.10. Synthesis of this

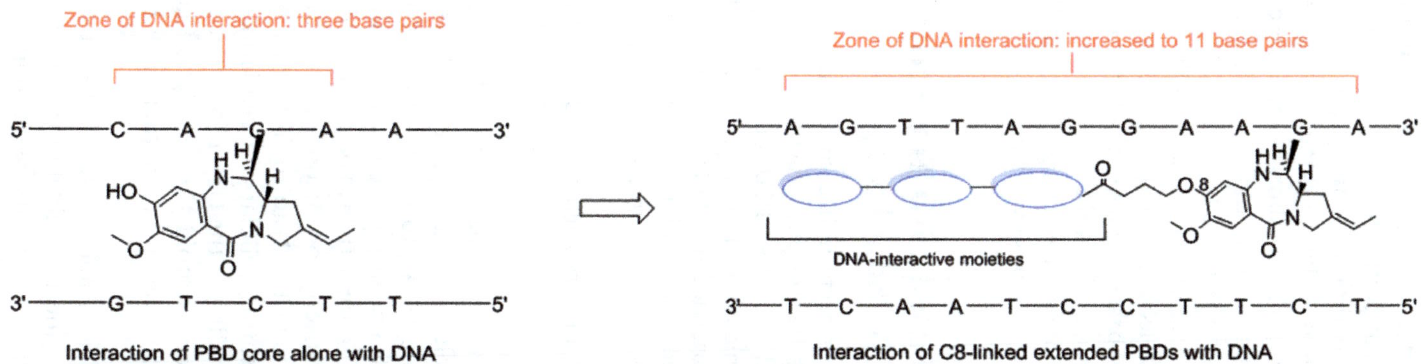

Figure 4.8 Diagram illustrating the rationale behind the development PBD-based molecules as transcription factor inhibitors.

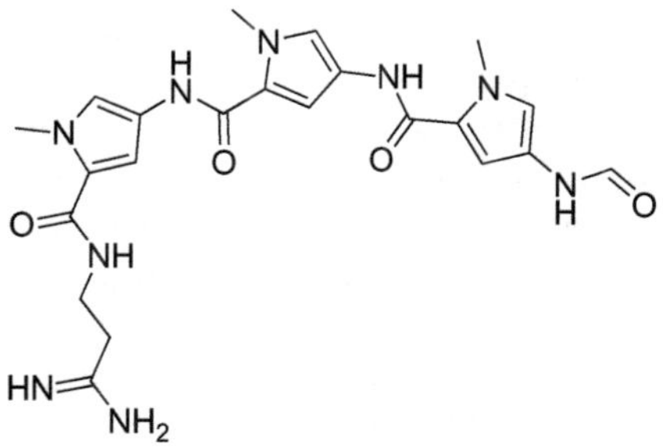

Figure 4.9 The structure of distamycin, a naturally occurring DNA-interactive polypyrrole agent from *Streptomyces* species that interacts non-covalently in the minor groove of DNA.

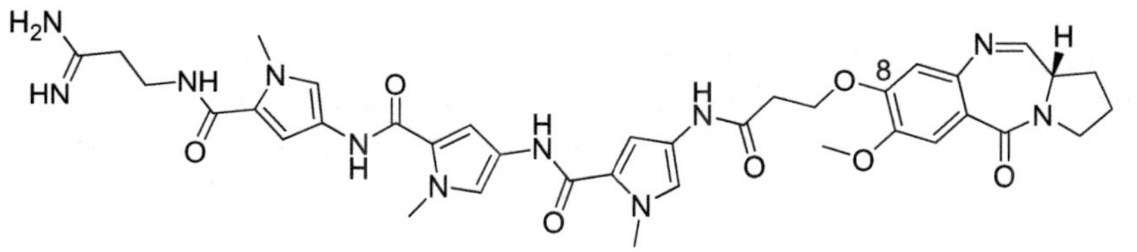

Figure 4.10 Structure of the first PBD–distamycin hybrid molecule synthesized by Baraldi and co-workers.[52]

compound involved initial construction of the PBD core, followed by conjugation of the distamycin moiety *via* a three-carbon linker to the C8-position of the molecule. Conjugation of chemical moieties at the C8-position of a PBD does not negatively affect its binding within the minor groove, and this is a commonly used attachment point for the synthesis of PBD hybrid molecules of many different types. Carbon linkers of variable length attached at the C8-position retain the good fit of the molecule in the minor groove by allowing greater flexibility and isohelicity compared with direct attachment of a moiety at the C8-posiion.[53] The Baraldi hybrid molecule was evaluated for cytotoxicity in the K562 cell line providing a concentration giving half the maximum inhibition (IC_{50}) value of 200 nM, a five times lower value than that obtained for a non-chemically conjugated mixture of the PBD core and distamycin. This demonstrated that both components of the Baraldi hybrid molecule contributed to the overall increase in cytotoxicity. It was assumed that this enhancement of biological activity was provided by the synergistic effect of the PBD core providing a covalently bonding foundation for the non-covalently AT-interacting distamycin moiety, and the overall increase of the footprint of the molecule to five base pairs. In an attempt to provide support for this hypothesis, PCR experiments were carried out in which both AT-rich and GC-rich DNA sequences were pre-incubated with the hybrid molecule to observe any inhibitory properties. An effect was observed for both sequences, in contrast to distamycin, which affected only the AT-rich sequence, indicating that the hybrid molecule could target both AT-and GC-rich regions of DNA. Interestingly, the unconjugated PBD core failed to appreciably reduce the PCR amplification of either GC- or AT-rich sequences.

Structure–activity relationships (SARs) between distamycin–PBD conjugates were then explored by lengthening the distamycin component to investigate the effect on activity. The addition of an extra pyrrole component to the C8-distamycin side chain was found to enhance cytotoxicity,[54] while shortening the chain resulted in inferior potency profiles. Lengthening the C8-distamycin component also enhanced the ability of the molecules to inhibit amplification of the AT-rich sequences compared with the GC-rich sequences in the PCR experiments. This was thought to be a direct consequence of the lengthening of the AT-recognizing C8-linked distamycin portion of the molecule. Finally, in DNA footprinting experiments, an increase in the length of footprints between distamycin alone and the distamycin–PBD hybrids indicated that the molecules were spanning a greater number of DNA bases than either the PBD or the distamycin components alone, and with an enhanced sequence-selectivity.

The ability of this distamycin–PBD hybrid to act as a transcription factor was then evaluated.[55] Using a filter-binding assay, the interaction of ^{32}P-radiolabelled SP-1 with DNA was initially measured. SP-1 is an essential transcription factor involved in the modulation of a large array of genes, including those involved in early development.[56] It preferentially binds to GC-rich DNA sequences, and its dysfunction has been implicated in the pathogenesis of several types of malignancies, including pancreatic and breast cancers.[57] After incubation of the DNA with the distamycin–PBD hybrid, inhibition of binding of SP-1 to its consensus sequences in K562 cells was clearly observed, which was not the case for either distamycin or the PBD core molecule alone. Further filter binding assays were carried out using purified NF-κB and ^{32}P-radiolabelled NF-κB DNA binding sequences, which demonstrated an inhibition of NF-κB binding by the distamycin–PBD hybrids.[58] These studies provided the first clear evidence that transcription factor binding can be influenced by distamycin–PBD molecules of this type, and paved the way for further investigations.

4.6 Distamycin-based PBD Transcription Factor Inhibitors: The Next Generation

The beginning of the 21st century heralded the development of synthetically modified versions of the original Baraldi distamycin–PBD hybrids, with a view to improving the potency and selectivity. Although early efforts proved unsuccessful, Lown and others[53,59,60] had some success by exploring the effects of modifying the distamycin C8-side chain. For example, the amidinoalkyl terminal unit was altered to a dimethylaminopropyl group as illustrated in Figure 4.11A. However, the potencies of these compounds remained largely equivalent to those of Baraldi hybrid molecules. The effect of the pyrrole components of the distamycin side chain on overall activity

A

B

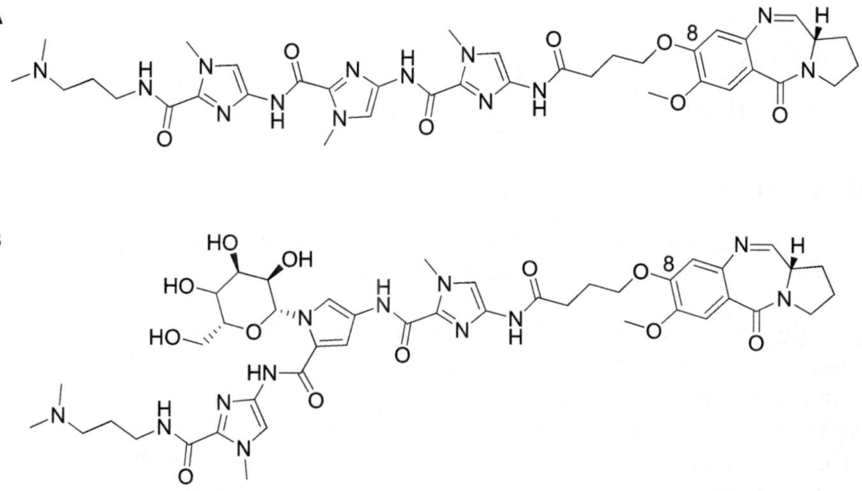

Figure 4.11 Examples of early C8-analogues of distamycin–PBD hybrids synthesised by Lown and others[53,59–61] to study SAR; (**A**) Molecules with a terminal dimethylaminopropyl group and/or imidazole rings replacing pyrroles, and (**B**) A glucuronide moiety conjugated to a pyrrole unit to enhance water solubility.

was also investigated. Partial or complete replacement of the side chain pyrroles with imidazole subunits (also see Figure 4.11A) was found to reduce or eliminate cellular cytotoxicity. Another avenue of exploration was to glycosylate a series of distamycin–PBD or polyimidazole–PBD conjugates to improve water solubility (Figure 4.11B), however these failed to show any improvement in activity compared with their non-glycosylated counterparts.[61] None of the analogues in Figure 4.11 was evaluated for effects on transcription factor activity.

A more successful approach involved the synthesis of a series of distamycin–PBD hybrids by Thurston and co-workers featuring a simplified C8-distamycin component.[53] For example, the amidinoalkyl terminal C8-functionality of the Baraldi hybrid was replaced with a simple methyl ester, with a view to enhancing the cellular and nuclear membrane penetration of compounds of this type. An extra carbon was also added to the linker between the C8-position of the PBD and the distamycin element to provide more flexibility (Figure 4.12).

The rationale for this alteration to the C8-linker was based upon a molecular modelling investigation, which indicated that a four-carbon linker might allow the hybrid molecule to fit within the DNA minor groove in a more effective manner. A small library of N-methyl PBD conjugates was then synthesised, with the number of pyrrole components conjugated to the C8-position of the PBD core ranging from one to six. A range of cytotoxicity and DNA-binding assays were then carried out on the library members. DNA thermal denaturation data confirmed that the DNA binding capacity of the hybrid molecules exceeded the additive binding stabilisation of the

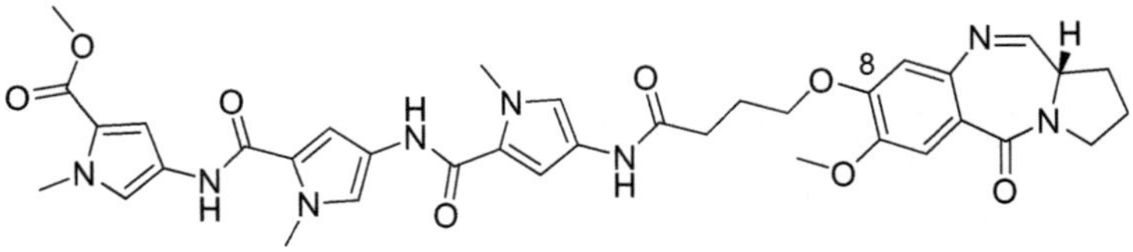

Figure 4.12 An example of a modified distamycin–PBD hybrid synthesised by Thurston and co-workers containing a simplified ester functionality at the terminus of the C8-side chain and an additional carbon in the linker immediately attached to the C8-position.

unconjugated components (*i.e.* the C8-side chain and the PBD core) tested in isolation. Interestingly, the highest thermal stabilisation of DNA was achieved by the analogue most similar in structure to Baraldi's initial three-pyrrole distamycin–PBD hybrid. This correlated with the results of an NCI cytotoxicity screen in which the analogue containing three pyrroles in the C8-side chain was the most cytotoxic of all analogues evaluated (*i.e.* $IC_{50} = 1.0$ μM). Overall, across the series of analogues evaluated, it was possible to correlate DNA-binding affinity with cytotoxicity. From a sequence-selectivity perspective, DNA footprinting studies using DNase detected unique binding sequences for each analogue evaluated. Individual binding motifs varied in length depending on the size of the C8-side chain of each hybrid molecule, but could be summarised generally as 5′-XGXW$_z$, where Z = 1–3, W = A or T bases, G = guanine and X = any base, but most commonly a purine.

In view of these encouraging results, a transcription inhibition assay was carried using RNA polymerase and an approximately 250 base pair nucleotide sequence. The polypyrrole–PBD hybrids inhibited transcription at much lower concentrations than the PBD core alone, indicating a higher affinity for DNA. The C8-side chain fragments alone failed to inhibit transcription even at high concentrations. The nature of the RNA fragments (*i.e.* "T-stops") produced by the polypyrrole–PBD hybrids were of interest when considering the potential transcription factor inhibitory capacity of the compounds. For example, the T-stops resulting from inhibition of RNA polymerase by the PBD core alone were numerous, indicating promiscuous binding of the core structure to a number of guanine-containing sites on the nucleotide. In contrast, the polypyrrole–PBD hybrids produced only approximately seven T-stops, which correlated closely to the binding sites observed in the footprinting assay. In summary, the transcription inhibition assay provided more evidence for the potency and DNA sequence-specificity of the polypyrrole–PBD hybrids, highlighting their potential use as inhibitors of transcription factor activity.

Given the encouraging potency and DNA sequence-selectivity characteristics displayed by the polypyrrole-PBD hybrids, their ability to inhibit transcription factor binding was further studied by Hartley and co-workers.[62] NF-Y is a transcription factor that is essential for the regulation of many eukaryotic genes, and is important in the control of cell cycle progression.

Closely resembling transcription factors found in yeasts,[63] this trimeric peptide binds selectively to gene promotors containing CCAAT motifs, thus controlling gene expression. However, due to the diversity of its role in gene control, the flanking sequences around this central CCAAT motif have also proved to be important. Thus, an ideal NF-Y binding site has been described as –C,G/A,G/C,C,A,A,T,C/G,A/G,C,A/C–.[64] When this is compared with the general binding motif 5′–XGXW$_z$ described for the polypyrrole–PBD hybrids described above, a considerable overlap is evident which indicated that NF-Y could be an ideal target for these compounds. As cancer cell gene expression has been shown to depend heavily on CCAAT box promotors and NF-Y overactivity (implicated in mutant p53-driven tumours), NF-Y inhibition was considered to represent a viable clinical target.[65]

Using GWL-78 (Figure 4.13), a member of the small library of polypyrrole–PBD hybrids, various assays were carried out to investigate the potential of the compound to inhibit the binding of NF-Y to its consensus CCAAT sequence. The first assay involved incubation of GWL-78 with immobilised oligonucleotide sequences containing the CCAAT sequence. The ligand-treated oligonucleotides were then exposed to cellular nuclear extracts containing NF-Y. Binding of NF-Y was evaluated using an ELISA assay, which specifically recognized the presence of bound transcription factor protein. The results indicated that NF-Y binding was reduced by 48% after ligand treatment compared with non-treated controls, indicating a significant affinity of GWL-78 for the CCAAT transcription factor recognition sequence and an ability to block protein interaction. An electrophoretic mobility shift assay (EMSA) was then undertaken in which the specific binding of GWL-78 to the CCAAT promoter regions of several genes (including topoisomerase II, mdr1, cdc25, E2F1 and cyclin B1) was studied. A difference in binding affinity of GWL-78 was observed not only between different genes (*e.g.* cdc25 binding eliminated at 50 $\mu M\,l^{-1}$ *versus* E2F1 binding eliminated at 10 $\mu M\,l^{-1}$) but also between different promoters on the same genes. For example, the topoisomerase II gene has four distinct inverted CCAAT box (ICB) promoter regions to which NF-Y may bind, thereby directing gene expression. During the EMSA assay, it was observed that concentrations of GWL-78 below 5 $\mu M\,l^{-1}$ abolished binding at the ICB-1 and ICB-2 sites, whereas 10 $\mu M\,l^{-1}$ was required to abolish binding at ICB-3 site. Interestingly, no abolition of binding was observed at the ICB-4 site at concentrations of GWL-78 up to 50 $\mu M\,l^{-1}$. These results demonstrated the potential for PBD hybrids, such as GWL-78, to have differential effects on specific subtypes of transcription

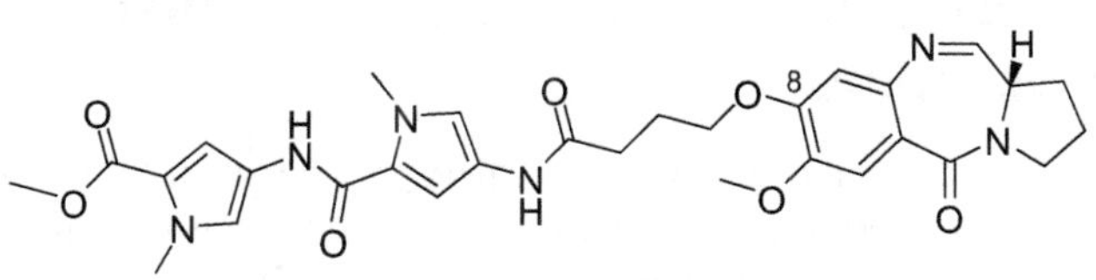

Figure 4.13 Structure of the polypyrrole–PBD hybrid GWL-78.

factor binding sites presumably based on DNA sequence-selectivity and binding affinity. This indicated that polypyrrole–PBD hybrids, such as GWL-78, might be further developed to target and modulate aberrant gene expression in malignant cells, while preserving homeostatic transcription factor activity in healthy surrounding tissues.

In order to test whether the results of the cell-free ELISA and EMSA assays could be reproduced in a cellular environment, a chromatin immunoprecipitation (ChIP) assay was carried out in NIH3T3, a fibroblast cell line. Before treatment with GWL-78, antibodies to NF-Y allowed extraction of DNA sequences bound to the transcription factor from nuclear extracts of the cells. After incubation of the cells with GWL-78, inhibition of NF-Y binding by the agent prevented recognition by the antibody and isolation of the corresponding DNA sequence. Promoters from each of the gene sequences investigated in the EMSA assay described above were evaluated for inhibition of NF-Y binding by GWL-78. Encouragingly, the ChIP assay results mirrored those obtained from the EMSA assay, with the topoisomerase IIa promoter sequence the most affected by GWL-78 and the cdc25 sequence the least affected.

Further assays provided additional evidence for the affinity of GWL-78 for the CCAAT DNA sequence. For example, DNA footprinting-based studies on a topoisomerase IIa gene fragment revealed that the molecule inhibited NF-Y binding at concentrations of 2.5 $\mu M\,l^{-1}$. Interestingly, the molecule was also shown to have the capacity to displace NF-Y, which was already bound to a CCAAT sequence. The effects of GWL-78 on the cell cycle were also evaluated using flow cytometry. Interestingly, although at higher doses cell cycle progression was inhibited, at lower doses it proceeded more rapidly. This paradoxical result was suggested to relate to the large number of genes, which may be regulated by NF-Y. At lower concentrations, GWL-78 may inhibit the expression of certain genes first (*e.g.* those with the most favourable binding sites, and/or more easily accessible promoter regions) associated with repression of cell cycle progression. In this way, cellular growth would be promoted. However, at higher levels of GWL-78, the ICB sequences which regulate the expression of genes associated with the promotion of cell cycle progression may be more affected (*e.g.* sites more difficult for the agent to access, and/or sites less favourable for binding), thus preventing NF-Y binding and halting cell growth.

4.7 Heterocyclic and Benzofusion-based PBD-based Transcription Factor Inhibitors: Further Improvement

The substantial data accumulated indicating that the polypyrrole–PBD hybrid GWL-78 can selectively inhibit binding of transcription factor NF-Y to DNA provided clear evidence that synthetic modification of the C8-side chain of distamycin–PBD hybrids can generate molecules with the ability to differentially inhibit transcription factors. The next step was to explore the SAR of the insertion of other heterocyclic and benzofused C8-side chains. One such investigation involved the replacement of the pyrrole subunits with imidazole or thiazole heterocycles to explore the effects on cytotoxic potency and DNA sequence-selectivity.[66] Initially, a small library of molecules was synthesised, in which the two pyrroles of GWL-78 were replaced partially or fully by either imidazole or thiazole components. Only one of the analogues synthesized (Figure 4.14), which had partial thiazole substitution, displayed an increase in activity compared with the parent molecule GWL-78.

This result, along with the lower activity of the imidazole series, confirmed the findings of Kumar and Lown[59] during a similar SAR study on the Baraldi distamycin–PBD hybrids. DNA denaturation assays were also carried out to determine whether the DNA-stabilising ability of the analogues had improved. With the exception of the partially thiazole-substituted analogue, all compounds evaluated stabilised DNA to a lesser extent than GWL-78, a result that correlated with those from the cytotoxicity studies.

However, although the introduction of imidazole and thiazole components lowered cytotoxicity, they did have some interesting effects on DNA sequence-selectivity. A DNA footprinting assay was carried out in which the compounds were incubated with the topoisomerase IIa promoter sequence. While GWL-78 showed no distinction in binding affinity between the ICB1 and ICB2 regions of the promoter sequence, two of the analogues had a binding preference for the ICB2 *versus* the ICB1 region. For example, a partially substituted thiazole analogue had a 25-times greater affinity for the ICB2 *versus* the ICB1 region, whereas a fully substituted imidazole analogue had only a five times greater affinity for the ICB2 region.

Although the substitution of the pyrrole element of GWL-78 had mixed effects on overall compound potency, the modification of DNA-binding affinity properties observed for some of the compounds in the study prompted further investigation.[67] A larger library of substituted C8-heterocyclic–PBD hybrids was synthesised in which the effects of imidazole (im) or thiazole (th) substitution on compounds containing a three-pyrrole C8-side chain were evaluated. Fourteen compounds were synthesised in total, featuring C8-side chains with partial or complete substitution of the pyrroles of the parent PBD hybrid. These compounds were evaluated for cytotoxicity in the NCI 60-cell line screen. Although all of the compounds tested had concentrations giving 50% of maximum lethality (LC_{50}) values in the low micromolar range,

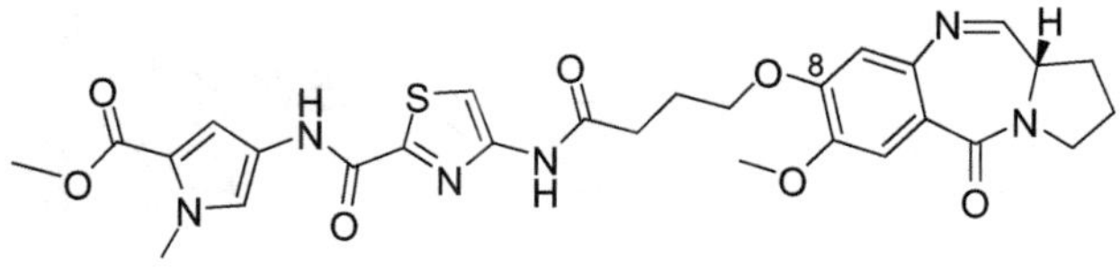

Figure 4.14 A GWL-78 analogue with a pyrrole subunit substituted for a thiazole moiety.

none was more active than the parent compound. This mirrored the cytotoxicity results for the substituted GWL-78 library. The fully substituted (*e.g.* im–im–im–PBD or th–th–th PBD) compounds were the least cytotoxic in the NCI screen (Figure 4.15), and were 15 times less potent than the parent compound. A DNA thermal denaturation assay provided similar results, although some monosubstituted hybrids displayed marginally improved DNA stabilisation profiles. DNA footprinting and EMSA assays were also carried out on a single monosubstituted library member, the Py–Py–im-PBD analogue. The DNA sequence used in these assays, the topoisomerase IIa promoter sequence, was the same as that used for the studies on the modified GWL-78 compounds described above. The Py–Py–im–PBD analogue showed a fivefold to tenfold preference for the ICB2 compared with the ICB1 binding site. However, it was unclear whether the original Py–Py–Py–PBD analogue also displayed these selectivity properties as, unfortunately, it was not evaluated in these studies, the relatively non-selective GWL-78 being used as a control instead.

Overall, the extensive studies on replacement of the pyrrole subunits of the original polypyrrole–PBD hybrids provided a mixed outcome, especially in terms of the disappointing potency of the new analogues produced. In many respects, this was unsurprising, as a similar trend had been observed by Kumar and Lown[59] during analogous modifications to Baraldi's distamycin–PBD hybrids. Essentially, it could be concluded from these studies that pyrrole units in the C8-side chain represented the optimal five-membered heterocycle building block for use in PBD-based transcription factor inhibitors. However, other researchers began to introduce alternative building blocks into the C8-side chain in an attempt to enhance DNA-specificity and potency. For example, Wang and co-workers synthesized the C8-linked PBD hybrids IN4CPBD and IN6CPBD (Figure 4.16) in which the traditional polypyrrole side chain had been changed to an indole unit linked to the C8-position of the PBD core *via* four- or six-carbon linkers, respectively.[68] These two analogues were synthesized as part of a larger library of PBD hybrids, which differed solely in terms of linker length. They were all found to be cytotoxic at the micromolar level in initial screens carried out in a melanoma cell line, and in the NCI tumour cell lines. IN6CPBD, featuring the six-carbon linker, was the most potent compound tested. Restriction enzyme inhibitory assays were also carried out, the results of which confirmed that both IN4CPBD and IN6CPBD interact efficiently with DNA.

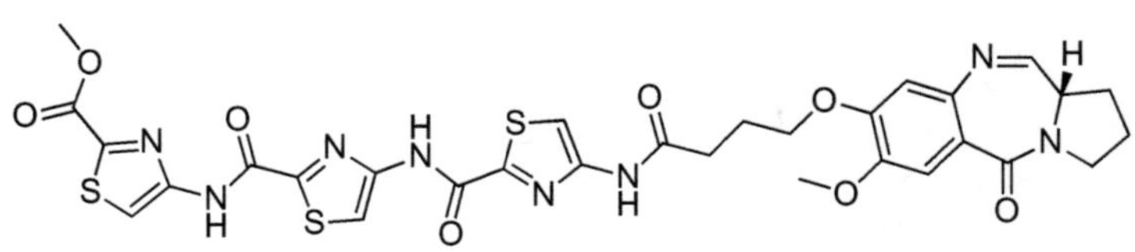

Figure 4.15 A modified C8-linked polypyrrole–PBD hybrid in which all of the pyrrole subunits have been substituted with thiazoles.

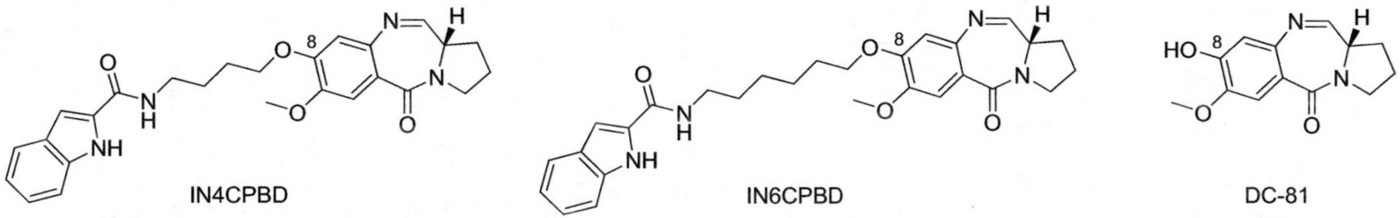

IN4CPBD IN6CPBD DC-81

Figure 4.16 Structures of the C8-linked PBD hybrids IN4CPBD and IN6CPBD synthesized by Wang and co-workers[68] in which the traditional five-membered pyrrole, imidazole or thiazole units in the C8-side chain were replaced with an indole heterocyclic unit. The linker between the C8-hydroxy position of the PBD and the heterocyclic unit was also increased to four and six methylene units, respectively. The structure of the natural product DC-81, which was used as a control in DNA-interactivity experiments, is also shown.

In addition, the results of flow cytometry studies indicated that the cell cycle was affected by both compounds, with an increase in sub-G1 material observed. Furthermore, apoptosis was detected in a series of assays involving the use of annexin V, and the measurement of loss of mitochondrial membrane potential.[69]

The cytotoxic potency and clear evidence of apoptosis upon cellular exposure to IN6CPBD prompted further investigation into its mechanism of action as a potential inhibitor and/or regulator of transcription factor activity.[70] Western blot studies demonstrated that raised cAMP levels upon exposure to IN6CPBD were found to activate a MAPK (mitogen activated protein kinase) cell-signalling pathway involving the c-Jun NH_2-terminal kinases (JNK). Regulation of the key cellular transcription factor AP-1 is controlled by activated JNK, which also plays a key role in the regulation of cellular apoptotic processes.[71] In support of this possibility, cellular AP-1 was found to increase upon exposure to one micromolar concentrations of IN6CPBD. Conversely, NF-κB activity decreased in response to exposure to IN6CPBD at similar concentration levels. Co-administration of an NF-κB inhibitor [*i.e.* Caffeic acid phenethyl ester (CAPE)] with IN6CPBD increased potency, suggesting that NF-κB inhibition may play an important role in the pro-apoptotic mechanism of action of this compound.

The alteration of the activation levels of both NF-κB and AP-1 in response to IN6CPBD treatment provided good evidence for transcription factor inhibition as a possible mechanism of action, although a definitive mechanism could not be fully elucidated. Interestingly, a recent study unrelated to IN6CPBD could help to provide an explanation for the results obtained by Wang and co-workers. A key link between the over-activity of NF-κB-mediated transcription and repression of the JNK pathway in multiple myeloma was recently reported.[72] These researchers found that, as NF-κB-mediated transcription increased, the activity of JNK-mediated signalling decreased due to the expression of a JNK-related repressor protein, growth arrest and DNA-damage-inducible protein 45b (GADD45b). The decoupling of this cross-talk between NF-κB and JNK through inhibition of GADD45b led to an increase in JNK activity, with a significant pro-apoptotic effect observed in the myeloma cell lines being studied. In the context of this potential linkage between NF-κB and JNK, a putative mechanism of action could emerge for IN6CPBD. Inhibition of NF-κB activity (as observed by Wang and colleagues) may abrogate the repression of the JNK pathway (AP-1 levels were observed to rise upon incubation with IN6CPBD), thus inducing cellular apoptosis. However, this proposed mechanism of action requires further experimental validation.

Investigations into the potential inhibition of metastatic processes in melanoma cell lines by IN4CPBD were also carried out but produced complex results.[73] Cellular motility of A375 melanoma cells was induced in culture using various concentrations of stromal cell-derived factor 1a (SDF-1a), a chemokine important for control of cellular motility. This motility was inhibited upon exposure of the cells to 1 nM IN4CPBD, and abolished at a concentration of 100 nM. Given the DNA-binding affinity of IN4CPBD, it is possible that the SDF-1a upregulation of cellular motility was abolished by this compound at the DNA transcriptional level. However, although Rho, Akt and extracellular signal–regulated kinase (ERK) 1/2 kinase activity was shown to be upregulated by SDF-1a binding, a single cell signalling pathway, or a transcription factor involved in modulation of gene expression in response to SDF-1a signalling could not be identified. This rendered the role of IN4CPBD in any specific sequence-selective inhibition difficult to identify. However, the study did provide some clues as to the mechanism of action at the downstream transcriptional level. For example, a reduction in the expression of MMP-9 (matrix metalloproteinase 9; an enzyme involved in tumour invasive processes) was observed upon exposure to IN4CPBD through quantitative PCR (qPCR), zymography and enzyme-linked immunosorbent assay (ELISA) studies. Expression of MMP-9 has been linked to SDF-1a signalling and has been shown to proceed *via* mitogen activated protein kinase (MAPK) and phosphoinositide 3-kinase (PI3K). Interestingly, NF-κB has been shown to mediate the expression of several genes modulated by the SDF-1a–PI3K signalling cascade.[74] As the closely related molecule IN6CPBD was shown to inhibit NF-κB activity,[70] it is reasonable to assume that IN4CPBD may act in a similar manner to inhibit the expression of MMP-P in melanoma cells. Similarly, VEGF (vascular endothelial growth factor), which plays an autocrine role in melanoma cell growth and proliferation, was shown to be modulated by IN4CPBD. This effect was proposed by Wang and coworkers[73] to be invoked through an inhibition of the PI3/Akt cell-signalling pathway. It is noteworthy that aberrant NF-κB activity has also been observed to be associated with VEGF expression in tumour cell lines,[75] indicating a potential inhibition of PI3K-activated NF-κB binding by IN4CPBD.

Regardless of the exact mechanism of transcription factor inhibition by these indole-containing PBD hybrids, an enhanced DNA sequence-selectively may translate into an improved safety profile in humans compared with traditional PBDs, which bind to DNA in a more promiscuous manner. This was demonstrated by further cell line experiments with IN6CPBD and by *in vivo* studies in mice.[76] For example, IN6CPBD displayed differential cytotoxicity toward individual cell lines when compared with a control PBD core. Using an annexin V assay, significant apoptosis was observed in B16 cells (melanoma) compared with healthy fibroblast cells. This differential induction of apoptosis was not observed for the unconjugated PBD core DC-81 (Figure 4.16) which affected both types of cells equally. *In vivo* experiments in mouse models with established lung metastases (typical of advanced melanoma pathogenesis) highlighted the antitumour efficacy and low toxicity of IN6CPBD. The compound decreased tumour burden in the mouse lungs by 30% more than similar doses of the unconjugated parent PBD DC-81. Crucially, the survival rate of IN6CPBD-treated mice also significantly increased compared with the DC-81-treated and control groups of mice. Interestingly, IN6CPBD produced significantly less toxicity than DC-81. Mice treated with DC-81 developed extensive hepatic and cardiac damage, which proved fatal to all mice treated at the highest dose of 10 mg kg^{-1}. In contrast, IN6CPBD had no effect on liver, heart or renal function even at the highest 10 mg kg^{-1} dose level. This study represented the first *in vivo* evaluation of a C8-conjugated PBD hybrid transcription factor inhibitor, and appeared to confirm the predictions of other researchers in this area that the enhanced specificity of DNA binding of molecules of this type should enhance biological activity while reducing toxicity due to the greater DNA sequence-selectivity with fewer DNA adducts formed per cell.

The development of the indole–PBD hybrid compounds IN4CPBD and IN6CPBD represented a significant step forward in the development of a PBD-based transcription factor inhibitor with clinical potential. In particular, the *in vivo* results supported the concept that sequence-selective covalent modification of DNA could lead to a low-toxicity therapy in contrast to current DNA-interactive cancer chemotherapy agents that form multiple DNA adducts across the genome which leads to significant toxicity. However, Wang and co-workers did not investigate the DNA sequence-selectivity of IN4CPBD and IN6CPBD, nor was the mechanism of action of these compounds fully elucidated. In addition, the molecules did not have improved potency compared with their predecessor C8-linked PBD hybrid molecules, with cellular IC_{50} values only in the micromolar range. This led to the next advance in the design of PBD-based transcription factor inhibitors, with molecules such as KMR-28-39 being synthesized.[77]

The development of KMR-28-39 originated in an attempt to alter the DNA sequence-selectivity of the polypyrrole/distamycin–PBD hybrids originally synthesised by Baraldi and Thurston. Although these original hybrid molecules possessed some specificity for guanine-containing sequences in the DNA minor groove due to the PBD unit, the AT-sequence specificity of the C8-polypyrrole/distamycin side chain rendered the molecules somewhat more selective for DNA sequences rich in adenine and thymine bases. While this enabled effective targeting of AT-rich sequences such as the ICB found in the promoter region of topoisomerase IIa,[62] binding to GC-rich sequences was less favoured. When the ubiquity of GC-rich sequences in gene promoter and enhancer sequences is taken into account (*e.g.* see review by Hapgood *et al.*[78]), the importance of the ability of any potential transcription factor inhibitor to target GC-rich sequences as well as AT-rich sequences (or a combination of the two) becomes clear. In order to develop a C8-linked PBD hybrid with greater GC-targeting ability, a library of biaryl building blocks featuring one heterocyclic and one carbocyclic component was prepared (Figure 4.17). These building blocks were then connected *via* amide bonds to produce potentially GC-interactive C8-side chains for connection to PBD core molecules. As well as different combinations of biaryl and five-membered heterocyclic components, these biaryl conjugates also contained a dimethylaminopropyl terminal moiety previously used in C8-linked PBD hybrids.[59,60]

The biaryl/heterocyclic building blocks were then evaluated for DNA sequence specificity using a fluorescent intercalation displacement (FID)

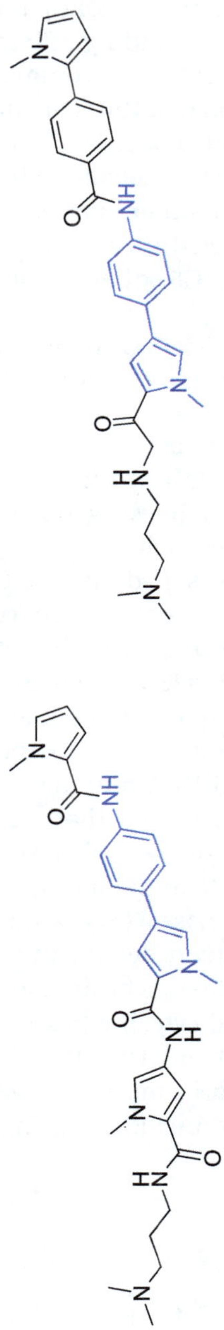

Figure 4.17 Examples of GC-interactive biaryl heterocycles that can be used as potential C8-PBD substituents. Both the examples shown contain a methyl pyrrole-benzenamine (MPB) subunit (highlighted in blue).

assay. The building blocks showed a clear preferential affinity for GC-rich compared with AT-rich sequences, the latter of which were preferred by polypyrrole and distamycin control molecules. Building blocks containing the biaryl methylpyrrole benzenamine (MPB) moiety (Figure 4.17) were identified as having the highest GC-binding affinity. With clear evidence of preferential GC-binding, building blocks containing MPB subunits were then conjugated to the C8-position of a PBD core molecule *via* a four-carbon linker, with a view to generating PBD-based GC-selective transcription factor inhibitors. Eight C8-linked biaryl or heterocyclic-PBD hybrids were synthesised, which differed from each other in terms of the number of biaryl or heterocyclic subunits in the C8-chain, the heterocyclic substituents used, and the overall orientation of the C8-subunits. An example of one of these hybrids is shown in Figure 4.18.

Evidence of GC sequence selectivity was demonstrated through a number of biophysical evaluations. For example, reversed-phase high performance liquid chromatography (HPLC) assays, using DNA hairpins as mimics of minor groove sequences, demonstrated effective binding to a GC-rich sequence by the biaryl/heterocyclic-PBD hybrid. Conversely, binding to an AT-rich sequence was shown to be much less favoured. Fluorescence resonance energy transfer (FRET)-based DNA melting assays were also carried out on both AT-rich and GC-rich sequences, and the results indicated that the latter were stabilised to a greater or at least equal extent compared with the AT-rich sequences. This contrasted to GWL-78, an earlier C8-linked polypyrrole–PBD hybrid, which is known to be AT-selective. Furthermore, DNA footprinting studies on a HexA nucleotide compared the sequence selectivity of KMR-28-39 with that of GWL-78.[113] KMR-28-39 gave footprints at both GC-rich and AT-rich sequences, and a general DNA-binding motif could be established for the molecule that was distinct from that for GWL-78. While GWL-78 bound favourably to sequences of N–G–N–N–W–W (where N = any base, G = guanine, W = A or T), KMR-28-39 bound sequences best described by N–G–N–W–N–S–S (where N = any base, G = guanine, W = A or T, S = G or C). This difference in sequence-selectivity demonstrated the increased tolerance of C8-linked MPB/heterocycle–PBD hybrids for sequences containing a higher proportion of guanine and cytosine bases.

Having established the sequence-selectivity of the compounds, the C8-linked MPB/heterocycle–PBD hybrid molecules were screened for cytotoxicity in a number of cell lines including primary chronic lymphocytic

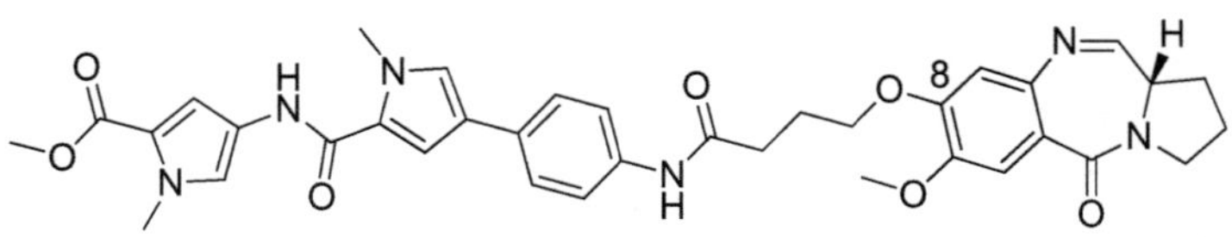

Figure 4.18 An example of a C8-linked biaryl/heterocyclic–PBD hybrid molecule containing a biaryl methylpyrrole benzenamine (MPB) moiety that provides a high affinity for GC-rich DNA sequences.

leukaemia (CLL) and healthy fibroblast cell lines. Low nanomolar LC_{50} values were observed for most compounds across all cell lines tested, with some (including KMR-28-39) even more potent (*e.g.* picomolar or femtomolar range) in certain cell lines. Interestingly, the non-tumour cell line WI-38 was 20–30-times less sensitive to the cytotoxic effect of most library members, indicating a potential selectivity window between tumour and non-tumour cells. It is noteworthy that GWL-78 was included in this cytotoxicity screen as a control molecule and proved to be more cytotoxic than previously reported (*i.e.* IC_{50} values in the nanomolar range, rather than the low micromolar range as observed in previous cytotoxicity screens). However, it was clear from these studies that the MPB/heterocycle–PBD hybrids were the most potent C8-linked PBD hybrid molecules synthesised to date in terms of their cytotoxicity in tumour cell lines.

An evaluation of SARs for the C8-side chain of the most active members of the library, KMR-28-35 and KMR-28-39 (Figure 4.19) provided some clues regarding the high cytotoxic potency of these molecules compared with previously synthesized PBD-based transcription factor inhibitors. For example, the positioning of a five-membered heterocycle immediately after the C8 four-carbon linker is common to both molecules, as is the positioning of the MPB unit at the end of the C8-side chain. Both molecules also feature a terminal pyrrole-methyl ester, a structural feature in common with the C8-linked polypyrrole–PBD hybrids synthesised by Wells and co-workers.[53] Although the structural features of the Wells and Rahman PBD hybrid libraries overlap to some extent, it is the differences in their DNA sequence-selectivity, potentially due to inclusion of the MPB unit in the C8-side chain of the KMR series (*e.g.* KMR-28-35 and KMR-28-39) that may account for their more potent cytotoxicity. In both the FRET-based DNA stabilisation and DNA footprinting assays described above, KMR-28-35 and KMR-28-39 demonstrated an ability to target mixed AT/GC sequences, contrasting with the properties of the AT-selective GWL-78, and the less cytotoxic members of the C8-linked MPB/heterocycle–PBD hybrid library, which demonstrated, in some cases, a higher GC sequence-selectivity. Overall, these results indicate that, for molecules of this type, a higher sequence-selectivity for a subset of the four DNA bases (*i.e.* AT- or GC-rich motifs) may not necessarily translate into improved cytotoxicity and that the ability to target motifs containing a mix of GC/AT bases may be more desirable.

Given this difference in DNA sequence selectivity, the effects of KMR-28-39 on the activity of a large range of transcription factors was evaluated in the HeLa cell line using a proprietary transcription factor activation profiling array assay (Signosis Inc). The activity of a number of transcription factors including nuclear factor of activated T-cells (NFAT), early growth response (EGR), NF-κB, small and mothers against decapentaplegic (SMAD) and octamer-binding transcription factor 4 (OCT-4) was shown to decrease by over 30% upon exposure of the cells to a 10 μM concentration of the agent. While the role of NF-κB in tumour cells and the development of cancer has been discussed above, over-activity of all of the other transcription factors

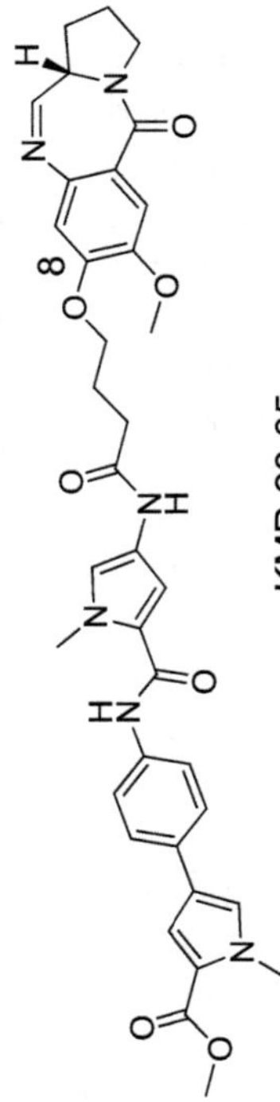

KMR-28-35

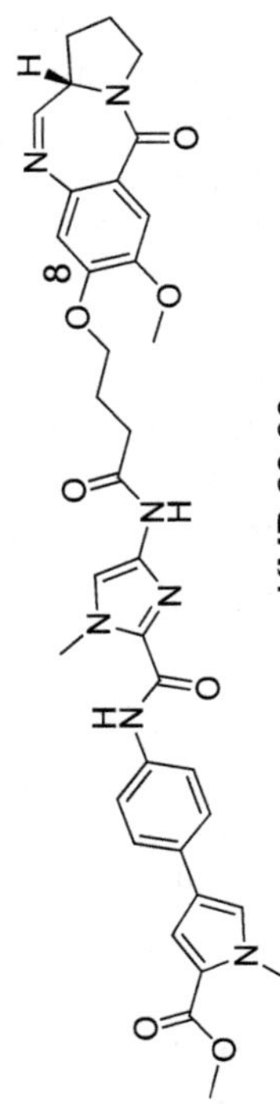

KMR-28-39

Figure 4.19 Structures of the C8-linked PBD hybrid molecules KMR-28-35 and KMR-28-39, both of which contain a biaryl methylpyrrole benzenamine (MPB) unit at the terminus of the C8-side chain. However, the molecules differ in that KMR-28-35 has an *N*-methylpyrrole unit at the C8-PBD end of the side chain, whereas KMR 28-39 has an *N*-methyl-imidazole unit at this position.

mentioned above (each heavily downregulated by KMR-28-39) has been implicated in carcinogenesis[79–82] across several different tumour cell types. Given the complex nature of transcription factor cross-talk and interactivity, it is currently unclear whether KMR-28-39 directly inhibits some or all of the individual transcription factors listed above, or indirectly inhibits one or more through modulation of another interdependent transcription factor. Although the mixed AT/GC sequence-selectivity of KMR-28-39 may allow direct inhibition within multiple transcription factor promoter/enhancer regions of genes, it is highly likely that inhibition of binding of one transcription factor will affect the activity levels of others. For example, it has been shown that NFAT and NF-κB interactivity may have some significance in the pathogenesis of lymphomas.[83]

As previous research on the distamycin–PBD and indole–PBD hybrids had indicated that they downregulate NF-κB, further studies of the effect of KMR-28-39 on NF-κB activity were conducted. Western blotting of phosphorylated IκB, a key kinase involved in the activation of the NF-κB complex, indicated a downregulation of activity upon exposure to KMR-28-39 at sub-nanomolar concentrations. The expression of BCL2 (a key anti-apoptotic protein) was also reduced, observed through immunohistochemical staining of KMR-28-39-treated tumours in a mouse model. As the expression of BCL2 is promoted by NF-κB, this provided further indirect evidence that KMR-28-39 can cause NF-κB downregulation.

After the promising *in vitro* cytotoxicity studies established for the MPB/heterocycle–PBD library, KMR-28-39 and the related KMR-28-35 were progressed to a tolerability evaluation (*i.e.* maximum tolerated dose) in mice. Both compounds were well tolerated at doses of 400 $\mu g\,kg^{-1}\,day^{-1}$ with few side effects observed other than slight weight loss. Consistent with the previous *in vivo* evaluation of indole-PBD hybrids,[76] this relative lack of systemic toxicity indicated that an improved safety profile for compounds of this type may be achievable due to their enhanced DNA sequence selectivity. As KMR-28-39 displayed a better tolerability profile compared with KMR-28-35 over repeat experiments, it was selected for human tumour xenograft studies in a mouse model based on human breast and pancreatic tumours. At a dose of 300 $\mu g\,kg^{-1}\,day^{-1}$ a significant reduction of tumour growth was observed in both models, which persisted for up to 14 days after the last dose, had been administered. KMR-28-39 is now in pre-clinical development.

The discovery of KMR-28-39, and the fact that it is now in pre-clinical development, owes much to the previous work of several research groups. For example, Baraldi and co-workers initially showed that sequence-selective DNA binding could be achieved through the synthesis of C8-linked distamycin–PBD hybrid molecules. The imidazole–PBD and polypyrrole–PBD hybrid analogues of Lown and co-workers and Thurston and co-workers, respectively, allowed the SARs of compounds of this type to be explored and provided key *in vitro* evidence of their ability to inhibit transcription factors. Next, the indole–PBD hybrids synthesised by Wang and co-workers demonstrated that a good safety profile could be achieved for PBD hybrids molecules in mouse studies. Finally, Rahman and co-workers synthesized KMR-28-39 and demonstrated that a higher cytotoxic potency is achievable *in vitro* while maintaining the ability to inhibit transcription factors. Therefore, the prospects for the future of C8-linked PBD transcription factor inhibitors based upon the structure of the original Baraldi distamycin hybrids is encouraging, and clinical studies are planned for KMR-28-39 and its close analogues in the near future.

4.8 Alternate PBD-hybrid Transcription Factor Inhibitors

In addition to the development of C8-linked monomeric PBD transcription factor inhibitors based on distamycin–PBD hybrids, other approaches to using PBD-based molecules for this purpose have been investigated. For example, a number of PBD dimers have been synthesized in which two PBD core molecules are joined together in head-to-head (*i.e.* C8–C8′-linked) or other arrangements (*e.g.* C2–C2′-, C7–C7′- or C2–C8′-linked) by carbon linkers of variable length (Figure 4.20).

The dimers featuring C2–C8′,[84] C2–C2′[85] and C7–C7′ linkages[86] between their PBD cores were found to be significantly less potent than their C8–C8′-linked counterparts. The C8–C8′-linked PBD dimers have been shown to form both intrastrand and interstrand DNA cross-links along with single-strand adducts, with the extent of each adduct type dependent on the C8–C8′-linker length and the availability and positions of suitable reactive guanine bases within a particular DNA sequence.[87] The interstrand cross-link is thought to be mainly responsible for the biological activity of the C8–C8′-linked PBD dimers, but may also be associated with their systemic toxicity.[88–90] Compared with traditional cross-linking agents, such as the nitrogen mustards (*e.g.* melphalan) and cisplatin, the DNA adducts formed have been shown to persist. This has been attributed to the good isohelical fit of PBD-based molecules within the DNA minor groove, which causes little distortion of the helix and thus does not attract the attention of DNA repair enzymes which can limit the activity of more traditional cross-linking agents.[89,91] Several dimeric C8–C8′-linked PBD molecules with various linker lengths and compositions have been synthesised,[87,92] with the most successful, SJG-136 (Figure 4.21)[93] reaching phase II clinical trials for the treatment of ovarian cancer and haematological malignancies.[94,95] SJG-136 is presently in a phase II clinical trial for canine cancer at the Royal Veterinary College (London).[96] Recently, PBD dimer analogues have also found use as cytotoxic payloads for ADCs, with a number of examples presently in clinical trials including at least one at the phase III level.[97,98]

The length of the DNA adducts formed by PBD dimers led to an investigation of the relationship between their DNA sequence-selectivity and the mechanism of their cytotoxicity. Initial studies showed that, depending on the length of their C8–C8′-linker, PBD dimers could have a preferential

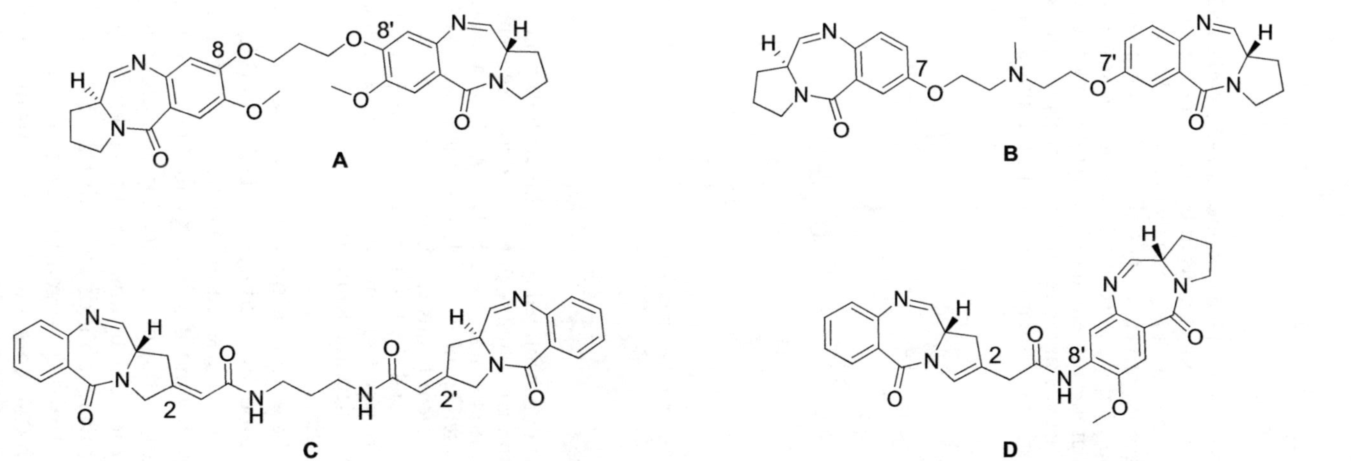

Figure 4.20 Examples of structures of PBD dimers: **A**, DSB-120, one of the first C8–C8′-linked PBD dimers; **B**, A C7–C7′-linked PBD dimer; **C**, A C2–C2′-linked PBD dimer; **D**, A C2–C8′-linked PBD dimer.

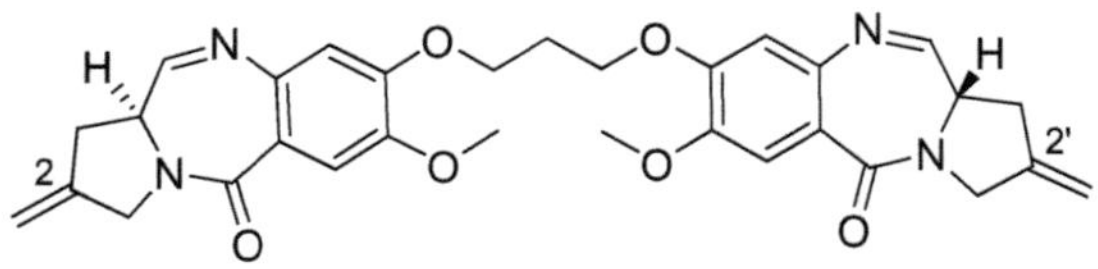

Figure 4.21 Structure of SJG-136, the best-known C8–C8′-linked PBD dimer, which reached phase II clinical trials in ovarian cancer and various haematological malignancies. It is presently in a phase II clinical trial for canine cancers at the Royal Veterinary College (London). For this PBD dimer, the addition of the exocyclic C2–C2′-methylene substituents to both PBD cores significantly enhanced the cytotoxicity and introduced *in vivo* antitumour activity in mouse and rat models in comparison to its non-C2–C2′-substituted parent molecule (DSB-120) which was inactive.

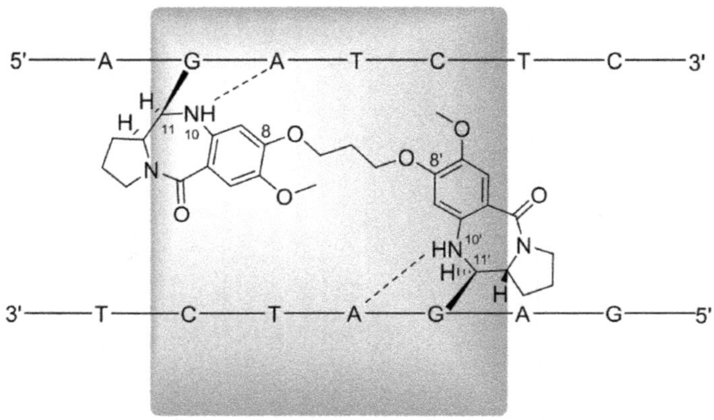

Figure 4.22 Model of the unsubstituted C8–C8′-linked PBD dimer DSB-120 forming an interstrand cross-link with a DNA duplex sequence (highlighted in grey). The symmetrical 5′-GATC/3′-GATC DNA motif bound by the agent is dictated by the two nucleophilic guanine bases on either strand that form covalent bonds *via* their C2-NH_2 functionalities to the C11–C11′-positions, and the two hydrogen bonds (shown as dashed lines) between the N10 and N10′ hydrogen atoms of each PBD core and the N3 atoms of the adjacent adenine bases.

selectivity for certain base pair sequences. Nuclear magnetic resonance (NMR) and molecular modelling studies of DSB-120 (Figure 4.22), a forerunner of SJG-136, showed that each PBD component retained a preference for the 5′–X–G–A–3′ binding motif (where X could be any base, but preferably a purine). This led to an overall preference for binding to a symmetrical 5′-GATC/3′-GATC motif dictated by the two nucleophilic guanine bases on either strand forming covalent bonds *via* their C2-NH_2 functionalities to the C11–C11′-positions, and two hydrogen bonds between the N10- and N10′-hydrogen atoms of each PBD core and the N3-atoms of the adjacent adenine bases (Figure 4.22). Thus, DSB-120 interstrand cross-links have the potential to interfere with a six base pair sequence of DNA in a relatively selective manner.[89,99]

The potent cytotoxicity of SJG-136 prompted further investigation of its DNA sequence-selectivity, which was initially evaluated through a DNA footprinting technique using a 262 base pair oligonucleotide and DNase1.[86] Footprints were observed at both occurrences of the predicted 5′–Pu–GATC–Py–3′ binding site within the 262 base pair fragment (*e.g.* a similar sequence selectivity to that shown for DSB-120 in Figure 4.22) suggesting a high affinity for the agent at this motif. However, several other footprints were observed which have more-recently been interpreted as resulting from variations to the central two bases of this motif (*e.g.* 5′–Pu–GXXC–Py–3′; where XX = TA, AA or TT) and to other DNA adduct types (*i.e.* intrastrand cross-links and mono-adducts). An *in vitro* transcription stop assay analogous to that described above for the C8-conjugated polypyrrole–DNA hybrids strongly supported these results by producing RNA polymerase stop sites that correlated well with the DNA footprints.[53] The RNA polymerase stop sites were also concentration-dependent, only appearing at higher concentrations of SJG-136. Finally, further evidence for the sequence-selectivity of SJG-136 was provided through a restriction endonuclease inhibition assay. Upon incubation of the agent with plasmid DNA, inhibition of the enzyme occurred preferentially at restriction sites containing the 5′–GATC–3′ motif.

In the past, the mechanism of activity of the PBD dimers has been attributed to the inhibition of DNA processing enzymes, such as RNA polymerase[42] and DNA endonucleases.[43] SJG-136 has also been shown to increase the levels of γH2AX in cells, a phosphorylated histone protein associated with DNA double-strand breaks.[100] However, based on the results described above, it is clear that PBD dimers bind selectively to certain sequences of DNA, spanning a sufficient number of base pairs to potentially interfere with the interaction of transcription factors whose DNA recognition motifs contain the PBD dimer binding site. Preliminary HPLC studies of SJG-136 binding to short synthetic oligonucleotides containing various transcription factor recognition motifs have recently been reported.[101] Interestingly, the agent was shown to bind to some transcription factor binding motifs, such as NF-κB and STAT-3, with a higher affinity than to others (*e.g.* AP-1). Also, the results of preliminary western blot studies have indicated that STAT-3-dependant genes are downregulated when breast tumour cells in culture are treated with SJG-136.[102] This is a significant result, as STAT3, a member of the Jak/STAT transcription factor family, has been implicated in the pathogenesis of a number of tumour types.[103] Furthermore, an interdependence between STAT-3 and NF-κB signalling pathways in some tumour types has been observed.[104] If the preliminary observations of the dual STAT-3–NF-κB inhibitory properties of SJG-136 can be confirmed and more clearly defined, they could help provide an explanation for the potent *in vitro* cytotoxicity and *in vivo* antitumour activity of SJG-136.[105,106]

Some researchers have taken a more radical approach in an attempt to improve the transcription factor inhibitory properties of PBD-based compounds. For example, Zhilina and co-workers designed and synthesized a C8-conjugate of a PBD and a triplex helix forming oligonucleotide (TFO) joined through a long polyethylene glycol (PEG) linker. TFOs are oligonucleotide fragments capable of non-covalent binding to polypurine sequences within the major groove of DNA to form triplex helix structures which have been shown to inhibit the interaction of DNA regulatory proteins (including transcription factors) in the major groove.[107] However, *in vitro* cellular experiments have shown that the biological effects of TFOs are usually short-lived, as the overall distortion of the DNA double helix caused by the binding of the TFO is recognised by the cell as DNA damage. This leads to activation of the DNA repair mechanisms which excise the non-covalently bound TFO.[107,108] Interestingly, the introduction of DNA alkylating moieties (*e.g.* diaziridinylquinones or phenylacetate mustards) into TFOs, which enhances binding affinity, has been shown to reduce the vulnerability to DNA repair. A corresponding increase in the sequence selectivity of TFOs conjugated to alkylating moieties has also been noted.[109–111] These observations led to the design of the PBD–TFO conjugate (shown in Figure 4.23) by Zhilina and co-workers in which the TFO moiety is anchored in position in the major groove by being tethered through the PEG linker to the PBD unit which is covalently bound in the minor groove.[112] These researchers had previously targeted the expression of a gene responsible for a key tyrosine kinase often dysregulated in cancer, the human epidermal growth factor receptor 2 (HER-2)/neu gene, using a stand-alone TFO specific for the promotor sequence of this gene. However, with this non-covalent-binding TFO, the biological effect was brief. Therefore, they decided to design the covalently binding TFO–PEG–PBD conjugate to try to obtain a more prolonged effect on HER-2/neu expression.

Unfortunately, evaluation of the TFO–PEG–PBD conjugate in *in vitro* experiments produced disappointing results, with the cytotoxicity of the conjugate being much lower by over 200-fold than for the TFO alone (*i.e.* $IC_{50} = 3.2$ μM *versus* 0.015 μM, respectively). EMSA assays were also carried out based on HER-2/neu DNA sequences which had inosine bases substituted for guanines at the PBD binding site in order to estimate the contribution of the PBD component to overall activity. Inosine, a guanine analogue, lacks the exocyclic amine at the C2-position, and so is incapable of binding to a PBD. No affinity difference was observed for the TFO–PEG–PBD conjugate between the two sequences, indicating a negligible effect of the PBD component on overall binding.

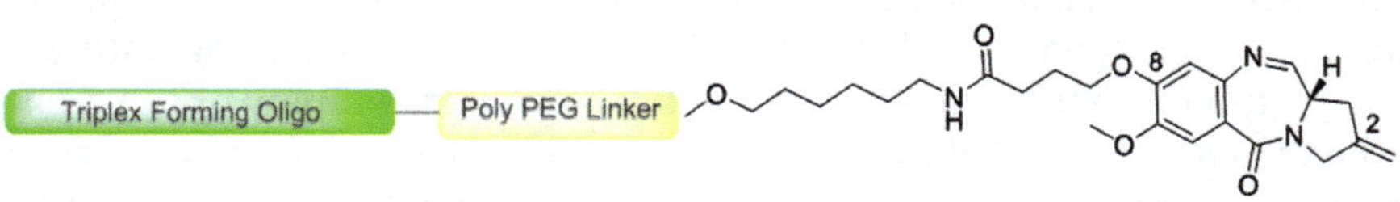

Figure 4.23 Schematic diagram of a C2-*exo*-methylene-substituted C8-conjugate of a PBD and a triple helix forming oligonucleotide (TFO) synthesized by Zhilina and co-workers.[112] A long polyethylene glycol (PEG) linker between the two components allows the PBD moiety to covalently bond in the minor groove and the TFO to reach the major groove where it binds non-covalently.

Therefore, although the concept of PBD-mediated stabilisation of TFO-based transcription factor inhibitors appears attractive, the practical utility seems to be limited based on the experiments carried out to date. In their publication[112] Zhilina and co-workers speculated that the relatively long PEG linker used to combine the minor groove (*i.e.* PBD) and major groove (*i.e.* TFO) binding components led to the lack of activity. Given the comparatively much shorter polypyrrole/distamycin–PBD conjugates synthesised by Thurston and Baraldi, which showed a correlation between lower cytotoxic potency and increased C8-linker length, the results for the TFO–PEG–PBD conjugate may not be too surprising. Moreover, the stabilisation of stand-alone TFO agents in the major groove by the conjugation of alkylating agents has been shown to be most improved by the inclusion of *bis*-alkylators, as mono-alkylating TFOs are more easily repaired.[110] Therefore, the design and synthesis of an analogue of the Zhilina TFO–PEG–PBD capable of forming cross-linked adducts in the minor groove (*i.e.* by attaching a PBD dimer instead of the monomeric PBD used by Zhilina and co-workers) may be worth investigating.

Many of the PBD-based hybrid transcription factor inhibitors described above resulted from a targeted approach to increase DNA sequence-selectivity. However, in some cases the PBD conjugates were initially designed to induce cytotoxic effects by different mechanisms. For example, the C8-conjugated quinazolinone-PBD hybrids synthesised by Kamal and co-workers (Figure 4.24) provide a good example of how transcription factor inhibitory activity may be unexpectedly observed for a chemically-modified PBD molecule.[114] These molecules were originally synthesised as an example of "combination chemotherapy" in which chemotherapeutic agents with different mechanisms of action were chemically conjugated together to produce hybrid molecules. The concept behind this approach was to improve upon the previously accepted strategy of using several chemotherapeutic agents in combination. In this case, the anti-tubulin properties of modified quinazolinones[115,116] were associated with an alkylating PBD core in an attempt to improve *in vitro* cytotoxicity and *in vivo* antitumour activity.

Using DNA thermal denaturation assays, it was initially found that incubation of these compounds with calf thymus DNA moderately increased its melting temperature, indicating that the PBD core of the hybrids was covalently interacting with the DNA. Sub-micromolar cytotoxicity was observed for several members of the compound library in cellular screens, prompting further investigation of their mechanism of action. Although an inhibitory

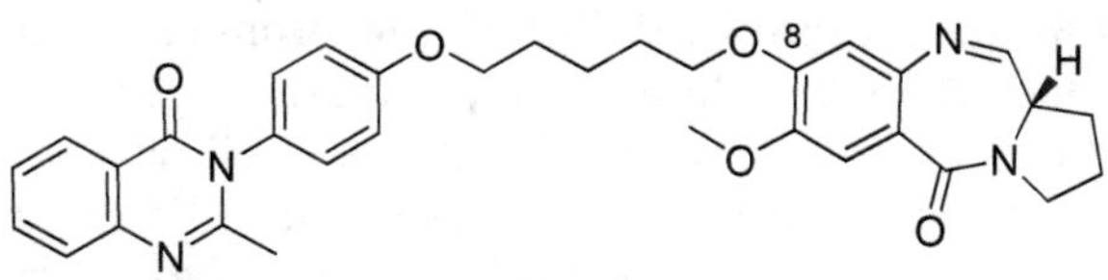

Figure 4.24 An example of a C8-quinazolinone–PBD hybrid synthesized by Kamal and co-workers.[114]

effect of the compounds on cell cycle progression was observed, and an upregulation of p53 noted, no investigation into potential anti-tubulin effects was made. As a result, it is difficult to determine the overall contribution of the quinazolinone moiety to the biological activity of the molecules. More interestingly from the perspective of this chapter, the molecules appeared to have an effect on the activity of NF-κB when introduced to the A375 melanoma cell line. In particular, western blot studies revealed a cellular decrease in p65, a key subunit of the NF-κB transcription factor, upon exposure to the quinazolinone–PBD hybrid. A corresponding decrease in BCL-2, an anti-apoptotic protein regulated by NF-κB binding, was also detected upon exposure to the conjugates, providing further evidence for a downregulation of NF-κB activity. It is unclear whether covalent interaction of these compounds with promoter sequences to which NF-κB binds was responsible for these effects, or whether NF-κB was inhibited in a more indirect manner. Other types of DNA-binding assays (*e.g.* DNA footprinting) could help to provide more insight into the precise mechanism of the transcription factor inhibitory ability of such hybrid molecules.

4.9 Conclusion

Historically, simple PBDs were considered to be traditional DNA-alkylating agents with characteristics such as potent cytotoxicity, promiscuous DNA binding and undesirable *in vivo* side effects. However, beginning with the conjugation of distamycin to the C8-position of a PBD core by Baraldi and co-workers, decades of research, including the synthesis of novel C8-conjugates, has started to alter that perception. Although the extraordinary length of DNA within the human genome means that no type of DNA-binding molecule is likely to be 100% sequence-specific, a significant degree of sequence-selectivity has been achieved for the latest generations of C8-linker PBD conjugates which, due to their length and the sequence-selective nature of newly designed building blocks, have quantifiable transcription factor inhibitory activity. In an age where sequence-selectivity has superseded cytotoxic potency as a critical design characteristic for novel DNA-interactive chemotherapeutics, the *in vivo* therapeutic index for some of these PBD-based transcription factor inhibitors is promising, especially in comparison with their unmodified PBD predecessors.

Overall, although the story of the development of PBD-based transcription factor inhibitors is a positive one, it remains incomplete. For example, further studies on the transcription-factor-inhibitory properties of PBD dimers such as SJG-136 are required in view of the promising preliminary data generated to date. This has become particularly urgent due to the progression of a number of PBD-containing ADCs to clinical trials in the past few years, some of which are at the phase III stage.[117] Clinicians working with these novel therapeutic entities will no doubt be interested to understand more about the precise mechanism of action of the cytotoxic PBD dimer payloads released from the antibody at the tumour site.

Regarding the C8-conjugated PBD hybrids, despite evidence that they can inhibit transcription factors such as NF-Y and NF-κB, it remains unclear whether this inhibition occurs directly through blocking the binding of the relevant transcription factor to DNA, or indirectly through modification of upstream activators or interconnected transcription factors. In reality, it may be a combination of both of these pathways. In the future, more focused molecular biology studies may help to unravel these potentially complex and intertwined mechanisms.

References

1. F. Spitz and E. E. Furlong, Transcription factors: from enhancer binding to developmental control, *Nat. Rev. Genet.*, 2012, **13**, 613–626.
2. W. S. Klug, Michael, C. Spencer, Palladino and Michael *Essentials of Genetics*, Benjamin Cummings, 7th edn, 2009.
3. D. S. Latchman, Transcription factors: an overview, *Int. J. Biochem. Cell Biol.*, 1997, **29**, 1305–1312.
4. J. Friedman and K. Kaestner, The Foxa family of transcription factors in development and metabolism, *Cell. Mol. Life Sci.*, 2006, **63**, 2317–2328.
5. P. Carlsson and M. Mahlapuu, Forkhead transcription factors: key players in development and metabolism, *Dev. Iiol.*, 2002, **250**, 1–23.
6. J. S. Rawlings, K. M. Rosler and D. A. Harrison, The JAK/STAT signaling pathway, *J. Cell Sci.*, 2004, **117**, 1281–1283.
7. C. Schindler and J. Darnell Jr., Transcriptional responses to polypeptide ligands: the JAK-STAT pathway, *Annu. Rev. Biochem.*, 1995, **64**, 621–652.
8. J. J. O'Shea, M. Gadina and R. D. Schreiber, Cytokine signaling in 2002: new surprises in the Jak/Stat pathway, *Cell*, 2002, **109**, S121–S131.
9. I. Costello, S. Nowotschin, X. Sun, A. W. Mould, A.-K. Hadjantonakis, E. K. Bikoff and E. J. Robertson, Lhx1 functions together with Otx2, Foxa2, and Ldb1 to govern anterior mesendoderm, node, and midline development, *Genes Dev.*, 2015, **29**, 2108–2122.
10. P. J. Mitchell and R. Tjian, Transcriptional regulation in mammalian cells by sequence-specific DNA binding proteins, *Science*, 1989, **245**, 371–378.
11. J. E. Darnell, Transcription factors as targets for cancer therapy, *Nat. Rev. Cancer*, 2002, **2**, 740–749.
12. J. Downward, Targeting RAS signalling pathways in cancer therapy, *Nat. Rev. Cancer*, 2003, **3**, 11–22.
13. S. Schubbert, K. Shannon and G. Bollag, Hyperactive Ras in developmental disorders and cancer, *Nat. Rev. Cancer*, 2007, **7**, 295–308.
14. M. S. Hayden and S. Ghosh, Shared principles in NF-κB signaling, *Cell*, 2008, **132**, 344–362.
15. S. Ghosh and M. Karin, Missing pieces in the NF-κB puzzle, *Cell*, 2002, **109**, S81–S96.
16. M. Barkett and T. D. Gilmore, Control of apoptosis by Rel/NF-kappaB transcription factors, *Oncogene*, 1999, **18**, 6910–6924.

17. A. S. Baldwin, Regulation of cell death and autophagy by IKK and NF-kappaB: critical mechanisms in immune function and cancer, *Immunol. Rev.*, 2012, **246**, 327–345.
18. U. Kordes, D. Krappmann, V. Heissmeyer, W. Ludwig and C. Scheidereit, Transcription factor NF-kappaB is constitutively activated in acute lymphoblastic leukemia cells, *Leukemia*, 2000, **14**, 399–402.
19. M. Nishio, T. Endo, N. Tsukada, J. Ohata, S. Kitada, J. C. Reed, N. J. Zvaifler and T. J. Kipps, Nurselike cells express BAFF and APRIL, which can promote survival of chronic lymphocytic leukemia cells via a paracrine pathway distinct from that of SDF-1α, *Blood*, 2005, **106**, 1012–1020.
20. M. L. Guzman, S. J. Neering, D. Upchurch, B. Grimes, D. S. Howard, D. A. Rizzieri, S. M. Luger and C. T. Jordan, Nuclear factor-κB is constitutively activated in primitive human acute myelogenous leukemia cells, *Blood*, 2001, **98**, 2301–2307.
21. V. L. Grandage, R. E. Gale, D. C. Linch and A. Khwaja, PI3-kinase/Akt is constitutively active in primary acute myeloid leukaemia cells and regulates survival and chemoresistance via NF-kappaB, Mapkinase and p53 pathways, *Leukemia*, 2005, **19**, 586–594.
22. Y. Maru, Molecular biology of chronic myeloid leukemia, *Cancer Sci.*, 2012, **103**, 1601–1610.
23. C. K. B. Ferrari, The biology of nuclear factor kappa beta (nfkb) in health and pathologY. Researchers, *World*, 2011, **2**, 53.
24. M. W. Mayo, C.-Y. Wang, P. C. Cogswell, K. S. Rogers-Graham, S. W. Lowe, C. J. Der and A. S. Baldwin, Requirement of NF-κB activation to suppress p53-independent apoptosis induced by oncogenic Ras, *Science*, 1997, **278**, 1812–1815.
25. D. S. Bassères, A. Ebbs, E. Levantini and A. S. Baldwin, Requirement of the NF-κB Subunit p65/RelA for K-Ras–Induced Lung Tumorigenesis, *Cancer Res.*, 2010, **70**, 3537–3546.
26. A. T. Baines, D. Xu and C. J. Der, Inhibition of Ras for cancer treatment: the search continues, *Future Med. Chem.*, 2011, **3**, 1787–1808.
27. P. Haluska, G. Dy and A. Adjei, Farnesyl transferase inhibitors as anticancer agents, *Eur. J. Cancer*, 2002, **38**, 1685–1700.
28. J. J. Fiordalisi, R. L. Johnson, C. A. Weinbaum, K. Sakabe, Z. Chen, P. J. Casey and A. D. Cox, High affinity for farnesyltransferase and alternative prenylation contribute individually to K-Ras4B resistance to farnesyltransferase inhibitors, *J. Biol. Chem.*, 2003, **278**, 41718–41727.
29. L. Lin, A. J. Sabnis, E. Chan, V. Olivas, L. Cade, E. Pazarentzos, S. Asthana, D. Neel, J. J. Yan and X. Lu, The Hippo effector YAP promotes resistance to RAF-and MEK-targeted cancer therapies, *Nat. Genet.*, 2015, **47**, 250–256.
30. R. B. Lobell, C. A. Omer, M. T. Abrams, H. G. Bhimnathwala, M. J. Brucker, C. A. Buser, J. P. Davide, S. Jane deSolms, C. J. Dinsmore and M. S. Ellis-Hutchings, Evaluation of farnesyl: protein transferase

and geranylgeranyl: protein transferase inhibitor combinations in preclinical models, *Cancer Res.*, 2001, **61**, 8758–8768.
31. F. R. Greten, M. C. Arkan, J. Bollrath, L.-C. Hsu, J. Goode, C. Miething, S. I. Göktuna, M. Neuenhahn, J. Fierer and S. Paxian, NF-κB is a negative regulator of IL-1β secretion as revealed by genetic and pharmacological inhibition of IKKβ, *Cell*, 2007, **130**, 918–931.
32. B. Gerratana, Biosynthesis, synthesis, and biological activities of pyrrolobenzodiazepines, *Med. Res. Rev.*, 2012, **32**, 254–293.
33. J. A. Hartley, The development of pyrrolobenzodiazepines as antitumour agents, *Expert Opin. Invest. Drugs*, 2011, **20**, 733–744.
34. E. Grunberg, H. Prince, E. Titsworth, G. Beskid and M. Tendler, Chemotherapeutic properties of anthramycin, *Chemotherapy*, 1966, **11**, 249–260.
35. K. M. Rahman, H. Rosado, J. B. Moreira, E. A. Feuerbaum, K. R. Fox, E. Stecher, P. W. Howard, S. J. Gregson, C. H. James, M. de la Fuente, D. E. Waldron, D. E. Thurston and P. W. Taylor, Antistaphylococcal activity of DNA-interactive pyrrolobenzodiazepine (PBD) dimers and PBD-biaryl conjugates, *J. Antimicrob. Chemother.*, 2012, **67**, 1683–1696.
36. C. Cargill, E. Bachmann and G. Zbinden, Effects of daunomycin and anthramycin on electrocardiogram and mitochondrial metabolism of the rat heart, *J. Natl. Cancer Inst.*, 1974, **53**, 481–486.
37. L. H. Hurley, T. Reck, D. E. Thurston, D. R. Langley, K. G. Holden, R. P. Hertzberg, J. R. Hoover, G. Gallagher Jr., L. F. Faucette and S. M. Mong, *et al.*, Pyrrolo[1,4]benzodiazepine antitumor antibiotics: relationship of DNA alkylation and sequence specificity to the biological activity of natural and synthetic compounds, *Chem. Res. Toxicol.*, 1988, **1**, 258–268.
38. S. Cheatham, A. Kook, L. H. Hurley, M. D. Barkley and W. Remers, One- and two-dimensional 1H NMR, fluorescence, and molecular modeling studies on the tomaymycin-d(ATGCAT)2 adduct. Evidence for two covalent adducts with opposite orientations and stereochemistries at the covalent linkage site, *J. Med. Chem.*, 1988, **31**, 583–590.
39. D. Graves, C. Pattaroni, B. Krishnan, J. Ostrander, L. Hurley and T. Krugh, The reaction of anthramycin with DNA. Proton and carbon nuclear magnetic resonance studies on the structure of the anthramycin-DNA adduct, *J. Biol. Chem.*, 1984, **259**, 8202–8209.
40. K. W. Kohn and C. L. Spears, Reaction of anthramycin with deoxyribonucleic acid, *J. Mol. Biol.*, 1970, **51**, 551–572.
41. K. M. Rahman, C. H. James, T. T. Bui, A. F. Drake and D. E. Thurston, Observation of a single-stranded DNA/pyrrolobenzodiazepine adduct, *J. Am. Chem. Soc.*, 2011, **133**, 19376–19385.
42. M. S. Puvvada, S. A. Forrow, J. A. Hartley, P. Stephenson, I. Gibson, T. C. Jenkins and D. E. Thurston, Inhibition of bacteriophage T7 RNA polymerase in vitro transcription by DNA-binding pyrrolo [2, 1-c][1, 4] benzodiazepines, *Biochemistry*, 1997, **36**, 2478–2484.

43. M. L. Kopka, C. Yoon, D. Goodsell, P. Pjura and R. E. Dickerson, The molecular origin of DNA-drug specificity in netropsin and distamycin, *Proc. Natl. Acad. Sci.*, 1985, **82**, 1376–1380.
44. R. P. Hertzberg, S. M. Hecht, V. L. Reynolds, I. J. Molineux and L. H. Hurley, DNA sequence specificity of the pyrrolo[1,4]benzodiazepine antitumor antibiotics. Methidiumpropyl-EDTA-iron(II) footprinting analysis of DNA binding sites for anthramycin and related drugs, *Biochemistry*, 1986, **25**, 1249–1258.
45. K. M. Rahman, H. Vassoler, C. H. James and D. E. Thurston, DNA Sequence Preference and Adduct Orientation of Pyrrolo[2,1-c][1,4]benzodiazepine Antitumor Agents, *ACS Med. Chem. Lett.*, 2010, **1**, 427–432.
46. P. Hardenbol, J. C. Wang and M. W. Van Dyke, Identification of preferred distamycin-DNA binding sites by the combinatorial method REPSA, *Bioconjugate Chem.*, 1997, **8**, 617–620.
47. C. Zimmer and U. Wähnert, Nonintercalating DNA-binding ligands: specificity of the interaction and their use as tools in biophysical, biochemical and biological investigations of the genetic material, *Prog. Biophys. Mol. Biol.*, 1986, **47**, 31–112.
48. U. H. Mortensen, T. Stevnsner, S. Krogh, K. Olesen, O. Westergaard and B. J. Bonven, Distamycin inhibition of topoisomerase I-DNA interaction: a mechanistic analysis, *Nucleic Acids Res.*, 1990, **18**, 1983–1989.
49. J. Woynarowski, M. McHugh, R. D. Sigmund and T. A. Beerman, Modulation of topoisomerase II catalytic activity by DNA minor groove binding agents distamycin, Hoechst 33258, and 4′, 6-diamidine-2-phenylindole, *Mol. Pharmacol.*, 1989, **35**, 177–182.
50. A. Taylor, K. A. Webster, T. A. Gustafson and L. Kedes, The anti-cancer agent distamycin A displaces essential transcription factors and selectively inhibits myogenic differentiation, *Mol. Cell. Biochem.*, 1997, **169**, 61–72.
51. M. Broggini, M. Ponti, S. Ottolenghi, M. D'Incalci, N. Mongelli and R. Mantovani, Distamycins inhibit the binding of OTF-1 and NFE-1 transfactors to their conserved DNA elements, *Nucleic Acids Res.*, 1989, **17**, 1051–1059.
52. P. G. Baraldi, B. Cacciari, A. Guiotto, A. Leoni, R. Romagnoli, G. Spalluto, N. Mongelli, P. W. Howard, D. E. Thurston and N. Bianchi, Design, synthesis and biological activity of a pyrrolo [2, 1-c][1, 4] benzodiazepine (PBD)-distamycin hybrid, *Bioorg. Med. Chem. Lett.*, 1998, **8**, 3019–3024.
53. G. Wells, C. R. Martin, P. W. Howard, Z. A. Sands, C. A. Laughton, A. Tiberghien, C. K. Woo, L. A. Masterson, M. J. Stephenson, J. A. Hartley, T. C. Jenkins, S. D. Shnyder, P. M. Loadman, M. J. Waring and D. E. Thurston, Design, synthesis, and biophysical and biological evaluation of a series of pyrrolobenzodiazepine-poly (N-methylpyrrole) conjugates, *J. Med. Chem.*, 2006, **49**, 5442–5461.
54. P. G. Baraldi, G. Balboni, B. Cacciari, A. Guiotto, S. Manfredini, R. Romagnoli, G. Spalluto, D. E. Thurston, P. W. Howard and N. Bianchi,

Synthesis, in vitro antiproliferative activity, and DNA-binding properties of hybrid molecules containing pyrrolo [2, 1-c][1, 4] benzodiazepine and minor-groove-binding oligopyrrole carriers, *J. Med. Chem.*, 1999, **42**, 5131–5141.
55. P. Baraldi, B. Cacciarit, A. Guiotto, R. Romagnoli, G. Spalluto, A. Leoni, N. Bianchi, G. Feriotto, C. Rutigliano and C. Mischiati, [2, 1-c][1, 4] benzodiazepine (PBD)-distamycin hybrid inhibits DNA binding to transcription factor Sp1, *Nucleosides, Nucleotides Nucleic Acids*, 2000, **19**, 1219–1229.
56. G. Suske, The Sp-family of transcription factors, *Gene*, 1999, **238**, 291–300.
57. S. Safe and M. Abdelrahim, Sp transcription factor family and its role in cancer, *Eur. J. Cancer*, 2005, **41**, 2438–2448.
58. M. Borgatti, C. Rutigliano, N. Bianchi, C. Mischiati, P. G. Baraldi, R. Romagnoli and R. Gambari, Inhibition of NF-kB/DNA Interactions and HIV-1 LTR Directed Transcription by Hybrid Molecules Containing Pyrrolo [2, 1-c][1, 4] Benzodiazepine (PBD) and Oligopyrrole Carriers, *Drug Dev. Res.*, 2003, **60**, 173–185.
59. R. Kumar and J. W. Lown, Synthesis and antitumor cytotoxicity evaluation of novel pyrrolo [2, 1-c][1, 4] benzodiazepine imidazole containing polyamide conjugates, *Oncol. Res. Featuring Preclin. Clin. Cancer Therap.*, 2002, **13**, 221–233.
60. R. Kumar, B. Reddy and J. W. Lown, Design and synthesis of novel pyrrolo [2, 1-c][1, 4] benzodiazepine-imidazole containing polyamide conjugates, *Heterocycl. Commun.*, 2002, **8**, 19–26.
61. R. Kumar and J. W. Lown, Design, synthesis and in vitro cytotoxicity studies of novel pyrrolo [2, 1][1, 4] benzodiazepine-glycosylated pyrrole and imidazole polyamide conjugates, *Org. Biomol. Chem.*, 2003, **1**, 3327–3342.
62. M. Kotecha, J. Kluza, G. Wells, C. C. O'Hare, C. Forni, R. Mantovani, P. W. Howard, P. Morris, D. E. Thurston and J. A. Hartley, Inhibition of DNA binding of the NF-Y transcription factor by the pyrrolobenzodiazepine-polyamide conjugate GWL-78, *Mol. Cancer Therap.*, 2008, **7**, 1319–1328.
63. X.-Y. Li, R. Mantovani, R. H. van Huijsduijnen, I. Andre, C. Benoist and D. Mathis, Evolutionary variation of the CCAAT-binding transcription factor NF-Y, *Nucleic Acids Res.*, 1992, **20**, 1087–1091.
64. R. Mantovani, The molecular biology of the CCAAT-binding factor NF-Y, *Gene*, 1999, **239**, 15–27.
65. D. Dolfini and R. Mantovani, Targeting the Y/CCAAT box in cancer: YB-1 (YBX1) or NF-Y&quest, *Cell Death Differ.*, 2013, **20**, 676–685.
66. F. Brucoli, R. M. Hawkins, C. H. James, G. Wells, T. C. Jenkins, T. Ellis, J. A. Hartley, P. W. Howard and D. E. Thurston, Novel C8-linked pyrrolobenzodiazepine (PBD)–heterocycle conjugates that recognize DNA sequences containing an inverted CCAAT box, *Bioorg. Med. Chem. Lett.*, 2011, **21**, 3780–3783.

67. F. Brucoli, R. M. Hawkins, C. H. James, P. J. Jackson, G. Wells, T. C. Jenkins, T. Ellis, M. Kotecha, D. Hochhauser, J. A. Hartley, P. W. Howard and D. E. Thurston, An extended pyrrolobenzodiazepine-polyamide conjugate with selectivity for a DNA sequence containing the ICB2 transcription factor binding site, *J. Med. Chem.*, 2013, **56**, 6339–6351.
68. J.-J. Wang, Y.-K. Shen, W.-P. Hu, M.-C. Hsieh, F.-L. Lin, M.-K. Hsu and M.-H. Hsu, Design, synthesis, and biological evaluation of pyrrolo [2, 1-c][1, 4] benzodiazepine and indole conjugates as anticancer agents, *J. Med. Chem.*, 2006, **49**, 1442–1449.
69. W.-P. Hu, H.-S. Yu, P.-J. Sung, F.-Y. Tsai, Y.-K. Shen, L.-S. Chang and J.-J. Wang, DC-81-indole conjugate agent induces mitochondria mediated apoptosis in human melanoma A375 cells, *Chem. Res. Toxicol.*, 2007, **20**, 905–912.
70. W.-P. Hu, F.-Y. Tsai, H.-S. Yu, P.-J. Sung, L.-S. Chang and J.-J. Wang, Induction of apoptosis by DC-81-indole conjugate agent through NF-κB and JNK/AP-1 pathway, *Chem. Res. Toxicol.*, 2008, **21**, 1330–1336.
71. R. J. Davis, Signal transduction by the JNK group of MAP kinases, *Cell*, 2000, **103**, 239–252.
72. L. Tornatore, A. Sandomenico, D. Raimondo, C. Low, A. Rocci, C. Tralau-Stewart, D. Capece, D. D'Andrea, M. Bua and E. Boyle, Cancer-selective targeting of the NF-κB survival pathway with GADD45β/MKK7 Inhibitors, *Cancer cell*, 2014, **26**, 495–508.
73. M.-C. Hsieh, W.-P. Hu, H.-S. Yu, W.-C. Wu, L.-S. Chang, Y.-H. Kao and J.-J. Wang, A DC-81-indole conjugate agent suppresses melanoma A375 cell migration partially via interrupting VEGF production and stromal cell-derived factor-1α-mediated signaling, *Toxicol. Appl. Pharmacol.*, 2011, **255**, 150–159.
74. S. R. Chinni, S. Sivalogan, Z. Dong, X. Deng, R. D. Bonfil and M. L. Cher, CXCL12/CXCR4 signaling activates Akt-1 and MMP-9 expression in prostate cancer cells: the role of bone microenvironment-associated CXCL12, *Prostate*, 2006, **66**, 32–48.
75. A. Shibata, T. Nagaya, T. Imai, H. Funahashi, A. Nakao and H. Seo, Inhibition of NF-κB activity decreases the VEGF mRNA expression in MDA-MB-231 breast cancer cells, *Breast Cancer Res. Treat.*, 2002, **73**, 237–243.
76. C.-H. Lee, W.-P. Hu, C.-H. Hong, H.-S. Yu, W.-T. Liao, C.-Y. Chen, Y.-L. Chen, B.-H. Chen, G.-S. Chen and J.-J. Wang, Pyrrolo [2, 1-c][1, 4] benzodiazepine and indole conjugate (IN6CPBD) has better efficacy and superior safety than the mother compound DC-81 in suppressing the growth of established melanoma in vivo, *Chem.-Biol. Interact.*, 2009, **180**, 360–367.
77. K. M. Rahman, P. J. Jackson, C. H. James, B. P. Basu, J. A. Hartley, M. de la Fuente, A. Schatzlein, M. Robson, R. B. Pedley and C. Pepper, GC-targeted C8-linked pyrrolobenzodiazepine–biaryl conjugates with femtomolar in vitro cytotoxicity and in vivo antitumor activity in mouse models, *J. Med. Chem.*, 2013, **56**, 2911–2935.

78. J. P. Hapgood, J. Riedemann and S. D. Scherer, Regulation of gene expression by gc - rich dna cis - elements, *Cell Biol. Int.*, 2001, **25**, 17–31.
79. M. R. Müller and A. Rao, NFAT, immunity and cancer: a transcription factor comes of age, *Nat. Rev. Immunol.*, 2010, **10**, 645–656.
80. T. Virolle, A. Krones-Herzig, V. Baron, G. De Gregorio, E. D. Adamson and D. Mercola, Egr1 promotes growth and survival of prostate cancer cells Identification of novel Egr1 target genes, *J. Biol. Chem.*, 2003, **278**, 11802–11810.
81. Y. Kang, W. He, S. Tulley, G. P. Gupta, I. Serganova, C.-R. Chen, K. Manova-Todorova, R. Blasberg, W. L. Gerald and J. Massagué, Breast cancer bone metastasis mediated by the Smad tumor suppressor pathway, *Proc. Natl. Acad. Sci. U. S. A.*, 2005, **102**, 13909–13914.
82. Y.-D. Wang, N. Cai, X. Wu, H. Cao, L. Xie and P. Zheng, OCT4 promotes tumorigenesis and inhibits apoptosis of cervical cancer cells by miR-125b/BAK1 pathway, *Cell Death Dis.*, 2013, **4**, e760.
83. L. V. Pham, A. T. Tamayo, L. C. Yoshimura, Y.-C. Lin-Lee and R. J. Ford, Constitutive NF-κB and NFAT activation in aggressive B-cell lymphomas synergistically activates the CD154 gene and maintains lymphoma cell survival, *Blood*, 2005, **106**, 3940–3947.
84. S. J. Gregson, P. W. Howard and D. E. Thurston, Synthesis of the first examples of A-C8/C-C2 amide-linked pyrrolo [2, 1-c][1, 4] benzodiazepine dimers, *Bioorg. Med. Chem. Lett.*, 2003, **13**, 2277–2280.
85. B. Reddy, Y. Damayanthi and J. Lown, Design, synthesis and in vitro cytotoxicity studies of novel pyrrolo [2, 1-c][1, 4] benzodiazepine (PBD)–polymade conjugates and 2, 2′-PBD dimers, *Anti-Cancer Drug Des.*, 2000, **15**, 225–238.
86. C. Martin, T. Ellis, C. J. McGurk, T. C. Jenkins, J. A. Hartley, M. J. Waring and D. E. Thurston, Sequence-selective interaction of the minor-groove interstrand cross-linking agent SJG-136 with naked and cellular DNA: footprinting and enzyme inhibition studies, *Biochemistry*, 2005, **44**, 4135–4147.
87. K. M. Rahman, C. H. James and D. E. Thurston, Effect of base sequence on the DNA cross-linking properties of pyrrolobenzodiazepine (PBD) dimers, *Nucleic Acids Res.*, 2011, **39**, 5800–5812.
88. D. S. Bose, A. S. Thompson, J. Ching, J. A. Hartley, M. D. Berardini, T. C. Jenkins, S. Neidle, L. H. Hurley and D. E. Thurston, Rational design of a highly efficient irreversible DNA interstrand cross-linking agent based on the pyrrolobenzodiazepine ring system, *J. Am. Chem. Soc.*, 1992, **114**, 4939–4941.
89. T. C. Jenkins, L. H. Hurley, S. Neidle and D. E. Thurston, Structure of a covalent DNA minor groove adduct with a pyrrolobenzodiazepine dimer: Evidence for sequence-specific interstrand crosslinking, *J. Med. Chem.*, 1994, **37**, 4529–4537.
90. M. Smellie, L. Kelland, D. Thurston, R. Souhami and J. Hartley, Cellular pharmacology of novel C8-linked anthramycin-based sequence-selective DNA minor groove cross-linking agents, *Br. J. Cancer*, 1994, **70**, 48.

91. J. A. Hartley, V. J. Spanswick, N. Brooks, P. H. Clingen, P. J. McHugh, D. Hochhauser, R. B. Pedley, L. R. Kelland, M. C. Alley and R. Schultz, SJG-136 (NSC 694501), a Novel Rationally Designed DNA Minor Groove Interstrand Cross-Linking Agent with Potent and Broad Spectrum Antitumor Activity Part 1: Cellular Pharmacology, In vitro and Initial In vivo Antitumor Activity, *Cancer Res.*, 2004, **64**, 6693–6699.
92. A. Kamal, P. M. M. Reddy, D. R. Reddy and E. Laxman, DNA binding potential and cytotoxicity of newly designed pyrrolobenzodiazepine dimers linked through a piperazine side-armed-alkane spacer, *Bioorg. Med. Chem.*, 2006, **14**, 385–394.
93. S. Gregson, P. Howard, D. Thurston, T. Jenkins and L. Kelland, Synthesis of a novel C2/C2′-exo unsaturated pyrrolobenzodiazepine cross-linking agent with remarkable DNA binding affinity and cytotoxicity, *Chem. Commun.*, 1999, 797–798.
94. Y. Y. Janjigian, W. Lee, M. G. Kris, V. A. Miller, L. M. Krug, C. G. Azzoli, E. Senturk, M. W. Calcutt and N. A. Rizvi, A phase I trial of SJG-136 (NSC# 694501) in advanced solid tumors, *Cancer Chemother. Pharmacol.*, 2010, **65**, 833–838.
95. I. Puzanov, W. Lee, A. P. Chen, M. W. Calcutt, D. L. Hachey, W. L. Vermeulen, V. J. Spanswick, C.-Y. Liao, J. A. Hartley and J. D. Berlin, Phase I pharmacokinetic and pharmacodynamic study of SJG-136, a novel DNA sequence selective minor groove cross-linking agent, in advanced solid tumors, *Clin. Cancer Res.*, 2011, **17**, 3794–3802.
96. M. Mellinas-Gomez, V. J. Spanswick, S. R. Paredes-Moscosso, M. Robson, R. B. Pedley, D. E. Thurston, S. J. Baines, A. Stell and J. A. Hartley, Activity of the DNA minor groove cross-linking agent SG2000 (SJG-136) against canine tumours, *BMC Vet. Res.*, 2015, **11**, 215.
97. M. S. K. Sutherland, R. B. Walter, S. C. Jeffrey, P. J. Burke, C. Yu, H. Kostner, I. Stone, M. C. Ryan, D. Sussman and R. P. Lyon, SGN-CD33A: a novel CD33-targeting antibody–drug conjugate using a pyrrolobenzodiazepine dimer is active in models of drug-resistant AML, *Blood*, 2013, **122**, 1455–1463.
98. K. Whiteman, C. Audette, A. Dandeneau, M. Ellis, N. Fishkin, L. Harvey, H. Johnson, Y. Kovtun, E. Maloney and M. Miller, Antibody-drug conjugates (ADCs) with a novel DNA-alkylating agent, DGN462, are highly potent in vitro and in vivo against human cancer models, *Cancer Res.*, 2014, **74**, 2644.
99. J. A. Mountzouris, J.-J. Wang, D. Thurston and L. H. Hurley, Comparison of a DSB-120 DNA interstrand cross-linked adduct with the corresponding bis-tomaymycin adduct: an example of a successful template-directed approach to drug design based upon the monoalkylating compound tomaymycin, *J. Med. Chem.*, 1994, **37**, 3132–3140.
100. J. Wu, P. H. Clingen, V. J. Spanswick, M. Mellinas-Gomez, T. Meyer, I. Puzanov, D. Jodrell, D. Hochhauser and J. A. Hartley, γ-H2AX foci formation as a pharmacodynamic marker of DNA damage produced by

DNA cross-linking agents: results from 2 phase I clinical trials of SJG-136 (SG2000), *Clin. Cancer Res.*, 2013, **19**, 721–730.
101. J. Mantaj, D. E. Thurston and K. M. Rahman, Interaction of SJG-136 with cognate sequences of oncogenic transcription factors, *Cancer Res.*, 2013, **73**, 2230.
102. J. Mantaj, D. E. Thurston and K. M. Rahman, Effect of the PBD dimer SJG-136 on expression of STAT3 dependent genes, *Cancer Res.*, 2015, **75**, 1986.
103. H. Yu and R. Jove, The STATs of cancer—new molecular targets come of age, *Nat. Rev. Cancer*, 2004, **4**, 97–105.
104. H. Yu, D. Pardoll and R. Jove, STATs in cancer inflammation and immunity: a leading role for STAT3, *Nat. Rev. Cancer*, 2009, **9**, 798–809.
105. J. A. Hartley, V. J. Spanswick, N. Brooks, P. H. Clingen, P. J. McHugh, D. Hochhauser, R. B. Pedley, L. R. Kelland, M. C. Alley, R. Schultz, M. G. Hollingshead, K. M. Schweikart, J. E. Tomaszewski, E. A. Sausville, S. J. Gregson, P. W. Howard and D. E. Thurston, SJG-136 (NSC 694501), a Novel Rationally Designed DNA Minor Groove Interstrand Cross-Linking Agent with Potent and Broad Spectrum Antitumor Activity. Part 1: Cellular Pharmacology, In vitro and Initial In vivo Antitumor Activity, 2004, **64**, 6693–6699.
106. M. C. Alley, M. G. Hollingshead, C. M. Pacula-Cox, W. R. Waud, J. A. Hartley, P. W. Howard, S. J. Gregson, D. E. Thurston and E. A. Sausville SJG-136 (NSC 694501), A Novel Rationally Designed DNA Minor Groove Interstrand Cross-Linking Agent with Potent and Broad Spectrum Antitumor Activity. Part 2: Efficacy Evaluations, 2004, **64**, 6700–6706.
107. M. P. Knauert and P. M. Glazer, Triplex forming oligonucleotides: sequence-specific tools for gene targeting, *Hum. Mol. Genet.*, 2001, **10**, 2243–2251.
108. J. Y. Chin and P. M. Glazer, Repair of DNA lesions associated with triplex-forming oligonucleotides, *Mol. Carcinogenesis*, 2009, **48**, 389–399.
109. S. L. Young, S. H. Krawczyk, M. D. Matteucci and J. J. Toole, Triple helix formation inhibits transcription elongation in vitro, *Proc. Natl. Acad. Sci.*, 1991, **88**, 10023–10026.
110. M. W. Reed, A. Wald and R. B. Meyer, Triplex-directed interstrand dna cross-linking by diaziridinylquinone-oligonucleotide conjugates, *J. Am. Chem. Soc.*, 1998, **120**, 9729–9734.
111. A. J. Ziemba, M. W. Reed, K. D. Raney, A. B. Byrd and S. W. Ebbinghaus, A bis-alkylating triplex forming oligonucleotide inhibits intracellular reporter gene expression and prevents triplex unwinding due to helicase activity, *Biochemistry*, 2003, **42**, 5013–5024.
112. Z. V. Zhilina, A. J. Ziemba, J. O. Trent, M. W. Reed, V. Gorn, Q. Zhou, W. Duan, L. Hurley and S. W. Ebbinghaus, Synthesis and evaluation of a triplex-forming oligonucleotide-pyrrolobenzodiazepine conjugate, *Bioconjugate Chem.*, 2004, **15**, 1182–1192.

113. A. J. Hampshire and K. Fox, Preferred binding sites for the bifunctional intercalator TANDEM determined using DNA fragments that contain every symmetrical hexanucleotide sequence, *Anal. Biochem.*, 2008, **2**, 298–303.
114. A. Kamal, E. V. Bharathi, M. J. Ramaiah, D. Dastagiri, J. S. Reddy, A. Viswanath, F. Sultana, S. Pushpavalli, M. Pal-Bhadra and H. K. Srivastava, Quinazolinone linked pyrrolo [2, 1-c][1, 4] benzodiazepine (PBD) conjugates: Design, synthesis and biological evaluation as potential anticancer agents, *Bioorg. Med. Chem.*, 2010, **18**, 526–542.
115. E. Hamel, C. M. Lin, J. Plowman, H.-K. Wang, K.-H. Lee and K. D. Paull, Antitumor 2, 3-dihydro-2-(aryl)-4 (1H)-quinazolinone derivatives: Interactions with tubulin, *Biochem. Pharmacol.*, 1996, **51**, 53–59.
116. M.-J. Hour, L.-J. Huang, S.-C. Kuo, Y. Xia, K. Bastow, Y. Nakanishi, E. Hamel and K.-H. Lee, 6-Alkylamino-and 2, 3-dihydro-3′-methoxy-2-phenyl-4-quinazolinones and related compounds: their synthesis, cytotoxicity, and inhibition of tubulin polymerization, *J. Med. Chem.*, 2000, **43**, 4479–4487.
117. J. Mantaj, P. J. M. Jackson, K. M. Rahman and D. E. Thurston, From Anthramycin to Pyrrolobenzodiazepine (PBD)-Containing Antibody–Drug Conjugates (ADCs), *Angew. Chem., Int. Ed.*, 2017, **56**, 462–488.

CHAPTER 5

Small Molecule Inhibitors of NF-κB and Their Therapeutic Potential in Leukaemia

CHRIS PEPPER* AND CHRIS FEGAN

Institute of Cancer & Genetics, Cardiff University, School of Medicine, Heath Park, Cardiff, CF14 4XN, UK
*Email: peppercj@cf.ac.uk

5.1 Introduction

Nuclear factor kappa B (NF-κB) is the collective name for a group of inducible homodimeric and heterodimeric transcription factors made up of members of the Rel (v-rel reticuloendotheliosis viral oncogene homologue) family of DNA-binding proteins. The NF-κB family of proteins consists of Rel A (p65), Rel B, c-Rel, p50 (NF-κB1) and p52 (NF-κB2).[1,2] In their inactive state, NF-κB proteins are trapped in the cytoplasm complexed with inhibitor of NF-κB (IκB) proteins.[2,3] NF-κB activation *via* the canonical signalling pathway relies on phosphorylation-dependent cleavage of p105, yielding p50, and subsequent degradation of IκB by sequential phosphorylation,[3] ubiquitination and proteasome-mediated proteolysis. This in turn releases the p50–p65 heterodimer, which translocates to the nucleus and binds to the promoters of NF-κB target genes.[1,3] NF-κB is capable of *trans*-activating more than 200 genes many of which are implicated in the control of cell survival and growth as well as the immune response and inflammation.[1] The principal NF-κB activation pathway involves IκB phosphorylation at serine residues 32 and 36

Drug Discovery Series No. 65
Small-molecule Transcription Factor Inhibitors in Oncology
Edited by Khondaker Miraz Rahman and David E. Thurston

Published by the Royal Society of Chemistry, www.rsc.org

by IκB kinase (IKK), which is made up of the IKKα, IKKβ and IKKγ subunits.[4] IKK can be activated by multiple intracellular and extracellular stimuli.[5]

The alternative NF-κB signalling pathway is triggered by members of the tumour necrosis family of cytokines *e.g.* B-cell activating factor of tumour necrosis factor family (BAFF) and CD40 ligand[6] and plays a crucial role in controlling B-cell survival and proliferation.[7] Finally, the atypical pathway is independent of IKK and can be activated by DNA damage and oxidative stress. The precise mechanism of atypical activation is not fully elucidated but appears to involve p38-activated casein kinase followed by tyrosine phosphorylation of IκB.[8] NF-κB activation is therefore highly complex and can be triggered by a diverse array of circumstances involving as many as 20 different protein kinases.[9–16]

5.2 The Role of NF-κB in Haematological Malignancies

The NF-κB family member c-Rel was the first subunit to be identified as a *bona fide* oncogene.[17] c-Rel is known for its potential to transform cells in culture, and it is frequently amplified in Hodgkin's lymphomas (HLs), diffuse large B-cell lymphomas (DLBCLs), and some follicular lymphomas.[18] More recently Rel A was shown to be an independent prognostic factor in chronic lymphocytic leukaemia patients, indicating that this subunit may also have oncogenic credentials.[19] In addition to the constitutive activation of NF-κB detected in many lymphoid and myeloid leukaemias, it has also been shown to be dysregulated in pre-malignant conditions like myelodysplastic syndromes (MDS), underlining its potential role in malignant transformation.[20]

5.2.1 NF-κB in Myeloid Malignancies

5.2.1.1 Myelodysplastic Syndromes

The myelodysplastic syndromes (MDS) are the most commonly diagnosed acquired bone marrow failure syndromes in adults.[21] The term MDS is used to describe a diverse group of myeloid disorders characterised by ineffective clonal haematopoiesis and abnormal "dysplastic" cell morphology, resulting in peripheral blood cytopenias.[22,23] In early MDS the peripheral cytopaenias can be explained by an increased apoptotic turnover of haematopoietic stem cells. In contrast, late-stage MDS often show a progressive accumulation of immature cells and the transformation to acute myeloid leukaemia.[20,24] Importantly, constitutive NF-κB activation has been shown to be a hallmark of high-risk MDS and precedes transformation to acute myeloid leukaemia (AML). Furthermore, NF-κB activation is not a feature of low-risk MDS, reinforcing its pathological significance in disease progression. In high-risk MDS nuclear translocation of Rel A is frequently detected in both $CD34^+$ haematopoietic stem cells and $CD33^+$ myeloid cells.[20] In keeping with the concept of Rel A playing a key role in transcriptional activation, the over expression of a number of NF-κB target genes, including B-cell lymphoma 2 (BCL2), B-cell leukaemia X long (BCL-X_L), inhibitor of apoptosis protein (IAP) and first apoptosis signal receptor-associated death domain-like interleukin-1β-converting enzyme-inhibitory protein (FLIP) have been correlated with the clinical progression of MDS.[25–27] Conversely, the expression of pro-apoptotic proteins of the Bcl-2 family like BCL-2 agonist/killer (Bak), BCL2-associated agonist of cell death (Bad), and B-cell leukaemia X short (Bcl-X_S) are good prognostic markers in MDS. Taken together, these data indicate that NF-κB-dependent repression of apoptosis is implicated in the transformation of MDS to AML.[25,26]

5.2.1.2 Acute Myeloid Leukaemia

Acute myeloid leukaemia (AML) represents a heterogeneous group of clonal malignancies arising from a putative leukaemic stem cells (LSC).[28–30] Several studies have demonstrated intrinsic activity of NF-κB in AML blasts and in isolated LSCs with a $CD34^+$–$CD38^-$–$CD123^+$ phenotype.[30,31] However, the mechanism for constitutive activation of NF-κB in these cells remains uncertain. One plausible mechanism is elevated IKK activity, resulting in the increased degradation of IκB.[32] However, constitutive overexpression of interleukin (IL)-1β, a potent NF-κB stimulator, has also been reported.[33] In concordance with high-risk MDS, the anti-apoptotic protein Bcl-2 is over expressed in AML with a particularly aggressive, apoptosis-resistant phenotype.[34] Interestingly, NF-κB activity was increased in all AML cytogenetic risk groups when compared with normal bone marrow samples,[32] but this was particularly evident in the poor cytogenetic risk group. This indicates that AML subtypes, which typically do not respond well to conventional therapy, might particularly benefit from treatment with NF-κB inhibitors.

5.2.1.3 Chronic Myeloid Leukaemia

Chronic myeloid leukaemia (CML) can perhaps be considered to represent the archetypal stem cell disease as it is characterised by the chromosomal translocation 9;22 which is manifest in all of the haematopoietic lineages. This reciprocal translocation results in the expression of the breakpoint cluster region protein–Abelson murine leukaemia viral oncogene homolog (Bcr–Abl) fusion protein, which exhibits constitutive kinase activity.[35] In turn, Bcr–Abl kinase signalling leads to the activation of a number of different survival pathways, including NF-κB.[36,37] Activation of NF-κB in CML is associated with increased translocation the Rel A subunit to the nucleus, which occurs in an IKK-independent manner.[36,37] NF-κB appears to be particularly evident in late-stage CML patients accompanied by the over expression of anti-apoptotic proteins like Bcl-X_L.[34,37]

5.2.2 NF-κB in Lymphoid Malignancies

5.2.2.1 *Acute Lymphoblastic Leukaemia*

The t(9;22) chromosomal translocation found in 95% of CML patients is also detectable in 25–30% of adult cases of B-precursor acute lymphocytic leukaemia (ALL) and approximately 5% of childhood ALL cases.[35] Just as is the case in CML, this leads to the transcription of the Bcr–Abl fusion protein, a serine–threonine kinase that activates NF-κB.[36] Blockade of NF-κB decreases the expression of the NF-κB targets c-MYC and BCL-X demonstrating that NF-κB is activated through the canonical IKK pathway.[37] To underscore the role of the *bcr–abl* translocation in the activation of NF-κB, Bcr–Abl$^+$ ALL blasts but not Bcr–Abl$^-$ tumour cells demonstrate NF-κB activation.[38] It is worthy of note that Bcr–Abl$^+$ patient samples do not exhibit dysregulated IKK activity and expression, indicative of an IKK-independent NF-κB activation process in these tumours. Consistent with this concept, Ras (Rat sarcoma virus gene homolog) mutations have been reported in ALL patients[39] and Ras can mediate IKK-independent translocation of Rel A to the nucleus.[36]

5.2.2.2 *Hodgkin's Lymphoma*

Hodgkin's lymphoma (HL) is defined by the presence of Hodgkin and Reed–Sternberg (HRS) cells derived from the clonal expansion of germinal-centre B-cells. These cells represent only a minority of the tumour bulk, the vast majority being B- and T-lymphocytes, macrophage and granulocytes. This creates a pro-inflammatory microenvironment, which leads to the activation of NF-κB *via* tumour necrosis factor (TNF) receptors expressed by the tumour.[40–42] In addition, up to 40% of HL patients are latently infected with Epstein–Barr virus (EBV), which can lead to the activation of NF-κB. The expression of the viral oncoprotein, latent membrane protein 1 (LMP1), can mimic CD40 signalling, thereby activating IKK.[43] Genetic mechanisms have also been implicated in driving constitutive NF-κB activity in HRS cells. Amplification of the REL locus is present in more than 50% of HL cases[43–45] and is associated with nuclear c-Rel staining. Biallelic deletions or frameshift mutations result in the activation of TNFAIP3 in 30–40% of HL cases.[46,47] Alternatively, frameshift or nonsense mutations in NF-κB inhibitor alpha (NFKBIA), result in truncated IκBα isoforms and occur in 5–20% of HL cases.[48]

5.2.2.3 *Non-Hodgkin's Lymphoma*

Non-Hodgkin lymphomas (NHLs) are a diverse group of blood cancers; the different sub-types of NHL vary significantly in their pathogenesis, and hence severity, from slow-growing to very aggressive. Similar to HL, genetic alterations that affect the activity and expression of NF-κB proteins have been linked to NHLs. For example, amplifications of the human c-REL locus, resulting in increased c-Rel expression, were found in a series of non-Hodgkin's B-cell lymphomas.[49] However, no direct correlation between c-Rel expression and the expression of anti-apoptotic NF-κB target genes was observed.[3] Increased c-Rel expression is implicated in malignant transformation because it was shown to transform primary chicken lymphoid cells *in vitro.*[50] In contrast to c-REL, amplifications and/or chromosomal rearrangements affecting the REL A locus, which encodes the NF-κB p65 subunit, are rare in NHLs.[51] In addition, no genetic alterations of the REL B locus have been described.[3] However, chromosomal rearrangements that affect the p52 (NF-κB2) locus at the 10q24 chromosomal region have been detected in T-cell non-Hodgkin's lymphomas, and also in chronic lymphocytic leukaemia and multiple myeloma.[52] All these rearrangements result in the deletion of C-terminal IκB-like sequences within the p100 and hence stimulate the constitutive transcriptional activity of p100-derived p52.

5.2.2.4 *Burkitt's Lymphoma*

Burkitt's lymphoma (BL) is an aggressive B-cell malignancy often associated with Epstein–Barr virus (EBV) infection. However, some sporadic cases appear to be independent of EBV infection.[53] Classically, BL is associated with the t(8;14)(q24;q32) translocation, in which the immunoglobulin promoter transcriptionally activates the c-MYC oncogene.[54] However, increased production of reactive oxygen species (ROS) and p38 mitogen activated protein kinase (MAPK) activation in EBV-positive BL can lead to NF-κB activation. In contrast, EBV-negative BL cells failed to produce ROS or to activate MAPK or NF-κB, indicating that BL can be classified into two distinct diseases involving different signalling pathways.[53]

5.2.2.5 *Diffuse Large B-cell Lymphoma*

NF-κB signalling has long been implicated in the pathogenesis of diffuse large B-cell lymphomas (DLBCL), which are thought to arise from normal, antigen-experienced, B-cells derived from the secondary lymphoid organs.[55] Microarray data indicated a classification of DLBCLs into germinal-centre-like and activated-B-cell (ABC)-like subtypes.[56,57] DLBCL cell lines with ABC-type signatures exhibit the constitutive activation of NF-κB target genes, as well as an increased sensitivity to cell death induction by NF-κB inhibition.[58] Most recently, whole-genome sequencing has led to the identification of three distinct DLBCL subtypes: the oxidative phosphorylation, activated B-cell (ABC) and host response (HR) subtypes.[59,60] The HR subtype exhibited a robust NF-κB target gene signature, while the activated B-cell (ABC)-like DLBCLs had a more restricted NF-κB target gene signature that resembles that found during the maturation of normal B-cells.[59] Amplification of the c-REL locus has been demonstrated in 23% of extranodal DLBCLs,[3] but NF-κB activation appears in the absence of c-REL amplification, indicating multiple causes of constitutive NF-κB activation.[3]

Gene profiling has revealed that NF-κB activation is common in NHLs but the precise mechanisms that drive this and the relevance of NF-κB activation to the pathology of these conditions remain to be characterised.[61] It is worthy of note that one report identified frequent dysregulation of the putative NF-κB target genes BCL2 and baculoviral inhibitor of apoptosis repeat-containing 5 (BIRC5)/survivin; elevated BIRC5/survivin correlated with an aggressive phenotype, while increased Bcl-2 expression correlated with less aggressive lymphomas.[61]

5.2.2.6 B-cell Chronic Lymphocytic Leukaemia

B-cell chronic lymphocytic leukaemia (CLL) is the most common leukaemia in the UK with an incidence of 3 in 100 000, equating to approximately 2000 newly diagnosed cases each year.[62] Previous studies have reported that CLL cells exhibit high constitutive NF-κB activation compared with normal B-lymphocytes.[63,64] More recently, we confirmed these findings but demonstrated a marked heterogeneity between patient samples in nuclear DNA binding of NF-κB.[65] Importantly, the constitutive expression of one Rel A, was correlated with *in vitro* susceptibility to spontaneous apoptosis and a number of clinically-relevant endpoints, including lymphocyte doubling time and tumour burden.[19]

The inter-patient variation in NF-κB activation seen in CLL is likely to result from a differential ability to respond to microenvironmental stimuli. In this context, the majority of these microenvironmental signals are likely to be encountered in the lymph nodes and or bone marrow.[66–69] It seems plausible that antigen binding to the B-cell receptor is an initiator event that triggers a number of additional NF-κB activation signals including CD40–CD40 ligand binding and interleukin-4; both derived from proximal activated T-cells.[63,64] Furthermore, we recently showed that *ex vivo* ligation of the B-cell receptor with anti-immunoglobulin M (IgM) induces one of two distinct cellular responses: CLL cells either induce NF-κB and enhance their *in vitro* survival or suppress NF-κB and become more susceptible to spontaneous apoptosis.[65] This adds weight to the argument that the signalling machinery is differentially hard-wired in individual CLL patients and this may be a key determinant in the pathology of this disease.[70,71] For instance, if a CLL cell responds to antigen by triggering NF-κB activation then this could lead to enhanced tumour cell survival, proliferation and clinical disease progression. Alternatively, if the CLL cell fails to induce NF-κB in response to B-cell receptor engagement then this may provide a self-limiting regulation of tumour cell numbers, resulting in a more stable and indolent form of the disease.

5.2.2.7 Multiple Myeloma

MM is a clonal B-cell malignancy characterised by the accumulation of clonal plasma cells in the bone marrow.[72] Thus far, MM remains an incurable disease. Recently, two studies revealed the expression of constitutively active NF-κB in freshly drawn bone marrow aspirates from 37 out of 37 MM patients.[73,74] Both studies demonstrated the retention of the NF-κB consensus sequence in nuclear extracts as well as translocation of the Rel A subunit to the nucleus in highly purified $CD138^+$ MM plasma cells. Both the canonical and noncanonical NF-κB pathways were activated[75] but the underlying mechanisms for expression of constitutively activated NF-κB in MM cells are poorly understood. Two non-mutually exclusive mechanisms could explain NF-κB activation in MM. Firstly, NF-κB could be activated in an autocrine manner; several NF-κB-inducing cytokines, including TNF, lymphotoxin (LT), IL-1, and receptor activator of NF-κB ligand (RANKL) are produced by MM cells and by the bone marrow microenvironment. Secondly, the IκB gene has been shown to be mutated in some MM patients.[76] Interestingly, the interplay between NF-κB and another transcription factor, signal transducer and activator of transcription 3 (STAT3), may contribute to the pathology of MM.[77] It is possible that NF-κB activation induces the expression of IL-6, which, in turn, activates STAT3.[73,74] Ras mutations occur late in the natural history of MM, leading to IL-6-independent growth.[78] The expression of the anti-apoptotic protein Bcl-X_L is induced by both STAT3 and NF-κB, and Bcl-X_L overexpression is frequently observed in MM cells.[77] This overexpression negatively correlates with response to chemotherapy, with only a minority of Bcl-X_L-expressing tumours responding to treatment.[79]

5.3 NF-κB and Drug Resistance

The activation of NF-κB has been linked to increased resistance to cellular apoptosis induced by a number of pharmacological stimuli, including a variety of DNA-damaging chemotherapeutic agents.[80–82] This enhanced activation of NF-κB, combined with the fact that many tumours show constitutive NF-κB activity prior to treatment, could represent a significant barrier to effective cancer treatment. The precise mechanisms of NF-κB-mediated drug resistance are not fully elucidated but recent studies have identified a consensus NF-κB binding site in the human multidrug resistance gene 1 (MDR1) and found that, at least *in vitro*, NF-κB transactivates the expression of MDR1.[83] This may result in the enhanced efflux of MDR1 substrates, including doxorubicin, etoposide, paclitaxel, tamoxifen, vincristine and topotecan.[84] In addition, NF-κB is known to regulate the transcription of a number of anti-apoptotic genes including BCL2, BCLX_L, MCL-1 (Myeloid Cell Leukaemia-1), BCL2 family member 1/antiapoptosis protein 1 (BFL1/A1) as well as inhibitor of apoptosis proteins (IAPs) and the caspase-8 regulating protein c-FLIP.[85–88] Therefore, a likely consequence of NF-κB inhibition would be to make cancer cells more susceptible to apoptosis and potentially enhance the therapeutic effects of conventional drugs.

Furthermore, we have shown that DNA binding of the NF-κB subunit Rel A was significantly higher in CLL samples derived from treated patients and was higher in treatment-naïve patients who went on to require therapy when compared with those who remained treatment-free.[19] This indicates that although exposure to chemotherapy increases Rel A DNA binding, basal Rel A can be used to prospectively risk-stratify patients. These data indicate that at least one NF-κB subunit is constitutively elevated in patients with more aggressive disease and is exacerbated by exposure to chemotherapy.

5.4 Therapeutic Targeting of NF-κB

The pivotal role of the NF-κB pathway in the inhibition of apoptosis, tumour promotion and progression, and the observation that NF-κB is constitutively activated in a large number of haematological malignancies, strongly indicate that NF-κB inhibitors would be useful in cancer therapy. Strategies for blocking NF-κB include upstream inhibition of NF-κB activation and direct competitive inhibitors of NF-κB.[89,90] Upstream inhibition strategies include the use of proteasome inhibitors (*e.g.* Bortezomib), IKK inhibitors [*e.g.* non-steroidal anti-inflammatory drugs (NSAIDs), sulfasalazine, curcumin and parthenolide analogues] and antioxidants (*e.g.* disulfiram and glutathione).[91,92] Direct targeting strategies include the use of cell-permeable peptide inhibitors and decoy oligodeoxynucleotides[93] and interfering with NF-κB mRNA using NF-κB antisense oligonucleotide.[94]

5.4.1 Pharmacological IKK Inhibitors

For many years the pharmaceutical industry has tried to develop selective IKK inhibitors. To date no selective IKKα inhibitors have been developed but, given its dual role in the two main pathways of NF-κB activation, IKKα represents an attractive target for cancer treatment. In contrast, a number of IKKβ inhibitors have been described and there are several reports confirming their anti-cancer activity both in cell lines and in primary cells through the induction of apoptosis.[95–97] One of the first IKKβ inhibitors synthesized was BAY117082 [(E)3-[(4-methylphenyl)-sulfonyl]-2-propenenitrile]. This compound is an irreversible inhibitor of IκBα phosphorylation, which blocks proteasomal degradation of IκBα and the subsequent translocation of NF-κB proteins to the nucleus.[98] BAY117082-induced cytotoxicity is associated with a rapid activation of the mitochondrial apoptotic pathway.[99] Bortezomib is a dipeptidyl boronic acid that specifically inhibits the 26S proteasome, the principal regulator of intracellular protein degradation.[100] This agent also blocks the degradation of IκBα, resulting in the cytoplasmic accumulation of NF-κB, but it remains unclear whether the therapeutic effects of bortezomib are dependent on NF-κB inhibition. It is approved as a single-agent treatment for multiple myeloma, and combination studies using various agents, including thalidomide and lenalidomide, are ongoing.[101,102] Obviously, bortezomib is now licensed in combination with other agents. Although the mechanisms of action of thalidomide and lenalidomide are complex, thalidomide has been shown to inhibit IKKβ.[103] Furthermore, thalidomide-induced apoptosis was associated with the repression of NF-κB target genes in multiple myeloma cells.[104]

5.4.2 Non-steroidal Anti-inflammatory Drugs

Several non-steroidal anti-inflammatory drugs (NSAIDs), including aspirin, ibuprofen, indomethacin and SDX-308, have been shown to suppress NF-κB activation by suppressing IKK activation and IκBα degradation.[105] Aspirin has been shown to induce apoptosis in a dose- and time-dependent manner in primary CLL cells, an effect not observed with other NSAIDs, at concentrations that inhibit cyclooxygenase. This indicates the involvement of cyclooxygenase-independent mechanisms in aspirin-induced cytotoxicity in CLL; a finding confirmed by our recent studies of novel aspirin analogues.[106] Furthermore, Sulfasalazine, a synthetic anti-inflammatory and immunosuppressive agent, has been shown to inhibit NF-κB activation *via* direct inhibition of IKKα and IKKβ and to induce apoptosis in primary CLL cells.[107] Most recently, the NSAIDs SDX-101 and SDX-308 have been shown to be very active against CLL cells. Their mechanism(s) of action are distinct; SDX-101 inhibits transcription of β-catenin whilst SDX-308 is an IKK inhibitor.[108]

5.4.3 Curcumin

Curcumin is a polyphenol derived from the plant *Curcuma longa*, and its biological effects are diverse, including anti-oxidant, anti-inflammatory, anti-angiogenic and anti-tumour activity. A number of recent studies have reported that curcumin induces apoptosis in CLL cells *via* the inhibition of STAT3, AKT and NF-κB.[109,110] Furthermore, combination studies with curcumin and the green tea extract epigallocatechin-3 gallate revealed synergy between these agents that was able to overcome stromal protection of CLL cells.[110] Resveratrol is another naturally occurring polyphenol compound. It is a potent inhibitor of NF-κB activation and NF-κB-dependent gene expression through its ability to inhibit IKK activity.[111]

5.4.4 Parthenolide Analogues

Parthenolide, a sesquiterpene lactone, has several known properties, including anti-inflammatory activity, inhibition of DNA synthesis and inhibition of cell proliferation in different cancer cell lines.[112–114] Furthermore, parthenolide is a potent inhibitor of NF-κB activation and has been shown to directly bind to IKK and to inhibit DNA binding by alkylating cysteine residues in the Rel A subunit.[115,116] It is a potent inducer of *in vitro* apoptosis but its usefulness as a drug is limited by poor water solubility. We recently reported that the water-soluble dimethylamino parthenolide analogue, LC-1, preferentially induces apoptosis in CLL cells and sensitizes them to the effects of fludarabine. Importantly, the apoptosis was preceded by a marked loss of NF-κB activity and sensitivity to LC-1 correlated with basal Rel A DNA binding.[117]

5.4.5 Competitive DNA Binding Inhibitors

A different approach to inhibit constitutive NF-κB activation is to interfere with NF-κB binding to DNA by means of phosphorothioate oligonucleotides that mimic the NF-κB consensus binding site. Such NF-κB-targeted decoy phosphorothioate oligonucleotides enhanced the apoptotic response to classical cytotoxic compounds.[118–120] However, NF-κB decoy oligonucleotides are relatively large and polar compounds, which negatively affects their cellular uptake and bioavailability.[5] An alternative approach is to use small-molecule inhibitors with sequence selectivity for NF-κB DNA binding motifs. We have recently shown that this is possible; inhibition of NF-κB was observed with femtomolar concentrations of agent.[121]

5.4.6 Targeting NF-κB Regulated Genes

Rather than direct targeting of NF-κB, an alternative approach is to inhibit specific NF-κB regulated genes. Some of the most promising of these are anti-Bcl-2 inhibitors designed to block protein–protein interactions. These so called Bcl-2 homology 3 BH3 mimetic agents are now starting to enter clinical trials; two such agents are ABT-737 and ABT-199.[122,123] As above, ABT-199 is now called Venetocalx and is licensed for use in CLL as a single agent and soon to be licensed as a combination agent with rituximab. These novel antagonist Bcl-2 inhibitors have shown high anti-tumour efficacy in CLL and other haematological cancers as well as demonstrating synergistic effects when used in conjunction with radiation and chemotherapy.[124,125] Other BH3 mimetics include WL-276, which was shown to inhibit the growth of drug-resistant prostate cancer,[126] and TW-37, which effectively attenuated the expression of NF-κB and its downstream target genes vascular endothelial growth factor (VEGF), cyclooxygenase 2 and BIRC5/survivin in pancreatic cancer.[127] Yet another novel small-molecule Bcl-2 inhibitor is (−)-Gossypol (AT-101). AT-101 was found to induce apoptosis in CLL[128] and was also recently shown to overcome stromal-mediated myeloid leukaemia cell differentiation protein gene (*Mcl-1*) induction and drug resistance.[129] Furthermore, Obatoclax (or GX15-070) has been shown to have single-agent activity against CLL and also shows potent synergy with chlorambucil and fludarabine.[130]

5.5 Conclusion

It is now clear that NF-κB plays a pivotal role in the initiation and maintenance of many haematological and solid cancers. It is also apparent that this transcription factor can modulate tumour cell responses to chemotherapeutic drugs. Therefore, targeting NF-κB may reverse drug resistance to standard chemotherapy. Although there are many examples of promising anti-NF-κB agents in pre-clinical development, the challenge is to translate this promise into clinical efficacy. Given that NF-κB plays a critical role in the activation of innate and adaptive immune responses, long-term use of pan-NF-κB inhibitors may not be desirable. However, it is now possible that we can target specific NF-κB subunits or individual NF-κB regulated genes. In this way, NF-κB inhibitors may prove to be very useful additions to the anti-cancer therapeutic arsenal.

References

1. B. B. Aggarwal, Nuclear factor-kappaB: the enemy within, *Cancer Cell*, 2004, **6**, 203–208.
2. M. S. Hayden and S. Ghosh, Signaling to NF-kappaB, *Genes Dev.*, 2004, **18**, 2195–2224.
3. M. Karin, Y. Cao, F. R. Greten and Z. W. Li, NF-kappaB in cancer: from innocent bystander to major culprit, *Nat. Rev. Cancer*, 2002, **2**, 301–310.
4. S. Ghosh and M. Karin, Missing pieces in the NF-kappaB puzzle, *Cell*, 2002, **109**, S81–S96.
5. M. Karin, Y. Yamamoto and Q. M. Wang, The IKK NF-kappa B system: a treasure trove for drug development, *Nat. Rev. Drug Discovery*, 2004, **3**, 17–26.
6. N. Yamagishi, J. Miyakoshi and H. Takebe, Enhanced radiosensitivity by inhibition of nuclear factor kappa B activation in human malignant glioma cells, *Int. J. Radiat. Biol.*, 1997, **72**, 157–162.
7. N. D. Perkins, Regulation of NF-kappaB by atypical activators and tumour suppressors, *Biochem. Soc. Trans.*, 2004, **32**, 936–939.
8. J. A. DiDonato, M. Hayakawa, D. M. Rothwarf, E. Zandi and M. Karin, A cytokine-responsive IkappaB kinase that activates the transcription factor NF-kappaB, *Nature*, 1997, **388**, 548–557.
9. F. Mercurio, H. Zhu, B. W. Murray, A. Shevchenko, B. L. Bennett, J. Li, D. B. Young, M. Barbosa, M. Mann, A. Manning and A. Rao, IKK-1 and IKK-2: cytokine-activated IkappaB kinases essential for NFkappaB activation, *Science*, 1997, **278**, 860–866.
10. E. Zandi, D. M. Rothwarf, M. Delhase, M. Hayakawa and M. Karin, The IkappaB kinase complex (IKK) contains two kinase subunits, IKKalpha and IKKbeta, necessary for IkappaB phosphorylation and NF-kappaB activation, *Cell*, 1997, **91**, 243–252.
11. C. C. Chen, Y. T. Sun, J. J. Chen and K. T. Chiu, TNF-alpha-induced cyclooxygenase-2 expression in human lung epithelial cells: involvement of the phospholipase C-gamma 2, protein kinase C-alpha, tyrosine kinase, NF-kappa B-inducing kinase, and I-kappa B kinase 1/2 pathway, *J. Immunol.*, 2000, **165**, 2719–2728.
12. A. C. Vertegaal, H. B. Kuiperij, S. Yamaoka, G. Courois, A. J. van der Eb and A. Zantema, Protein kinase C-alpha is an upstream activator of the IkappaB kinase complex in the TPA signal transduction pathway to NF-kappaB in U20S cells, *Cell Signal.*, 2000, **12**, 759–768.
13. Q. Zhao and F. S. Lee, Mitogen-activated protein kinase/ERK kinase kinases 2 and 3 activate nuclear factor-kappaB through IkappaB

kinase-alpha and IkappaB kinase-beta, *J. Biol. Chem.*, 1999, **274**, 8355–8358.
14. J. Yang, Y. Lin, Z. Guo, J. Cheng, J. Huang, L. Deng, W. Liao, Z. Chen, Z. Liu and B. Su, The essential role of MEKK3 in TNF-induced NfkappaB activation, *Nat. Immunol.*, 2001, **7**, 620–624.
15. J. B. Petro, S. M. Rahman, D. W. Ballard and W. N. Khan, Bruton's tyrosine kinase is required for activation of IkappaB kinase and nuclear factor kappaB in response to B cell receptor engagement, *J. Exp. Med.*, 2000, **191**, 1745–1754.
16. N. L. Malinin, M. P. Boldin, A. V. Kovalenko and D. Wallach, MAP3K-related kinase involved in NF-kappaB induction by TNF, CD95 and IL-1, *Nature*, 1997, **385**, 540–544.
17. T. D. Gilmore, Multiple mutations contribute to the oncogenicity of the retroviral oncoprotein v-Rel, *Oncogene*, 1999, **18**, 6925–6937.
18. T. D. Gilmore, D. Kalaitzidis, M. C. Liang and D. T. Starczynowski, The c-Rel transcription factor and B-cell proliferation: a deal with the devil, *Oncogene*, 2004, **23**, 2275–2286.
19. S. Hewamana, T. T. Lin, C. Rowntree, K. Karunanithi, G. Pratt, R. Hills, C. Fegan, P. Brennan and C. Pepper, Rel A is an independent biomarker of clinical outcome in chronic lymphocytic leukemia, *J. Clin. Oncol.*, 2009, **27**, 763–769.
20. T. Braun, G. Carvalho, A. Coquelle, M. C. Vozenin, P. Lepelley, F. Hirsch, J. J. Kiladjian, V. Ribrag, P. Fenaux and G. Kroemer, NF-{kappa}B constitutes a potential therapeutic target in high-risk myelodysplastic syndrome, *Blood*, 2006, **107**, 1156–1165.
21. M. A. Sekeres, The epidemiology of myelodysplastic syndromes, *Hematol. Oncol. Clin. North Am.*, 2010, **24**, 287–294.
22. J. W. Vardiman, Hematopathological concepts and controversies in the diagnosis and classification of myelodysplastic syndromes, *Am. Soc. Hematol. Educ. Program.*, 2006, 199–204.
23. D. P. Steensma and J. M. Bennett, The myelodysplastic syndromes: diagnosis and treatment, *Mayo Clin. Proc.*, 2006, **81**, 104–130.
24. G. Carvalho, C. Fabre, T. Braun, J. Grosjean, L. Ades, F. Agou, E. Tasdemir, S. Boehrer, A. Israel, M. Véron, P. Fenaux and G. Kroemer, Inhibition of NEMO, the regulatory subunit of the IKK complex, induces apoptosis in high-risk myelodysplastic syndrome and acute myeloid leukemia, *Oncogene*, 2007, **26**, 2299–2307.
25. M. Benesch, U. Platzbecker, J. Ward, H. J. Deeg and W. Leisenring, Expression of FLIP(Long) and FLIP(Short) in bone marrow mononuclear and CD34+ cells in patients with myelodysplastic syndrome: correlation with apoptosis, *Leukemia*, 2003, **17**, 2460–2466.
26. D. Boudard, C. Vasselon, M. F. Bertheas, J. Jaubert, C. Mounier, J. Reynaud, A. Viallet, S. Chautard, D. Guyotat and L. Campos, Expression and prognostic significance of Bcl-2 family proteins in myelodysplastic syndromes, *Am. J. Hematol.*, 2002, **70**, 115–125.

27. K. Yamamoto, S. Abe, Y. Nakagawa, K. Suzuki, M. Hasegawa, M. Inoue, M. Kurata, K. Hirokawa and M. Kitagawa, Expression of IAP family proteins in myelodysplastic syndromes transforming to overt leukemia, *Leuk. Res.*, 2004, **28**, 1203–1211.
28. Z. Estrov, S. Shishodia, S. Faderl, D. Harris, Q. Van, H. M. Kantarjian, M. Talpaz and B. B. Aggarwal, Resveratrol blocks interleukin-1beta-induced activation of the nuclear transcription factor NF-kappaB, inhibits proliferation, causes S-phase arrest, and induces apoptosis of acute myeloid leukemia cells, *Blood*, 2003, **102**, 987–995.
29. M. L. Guzman, C. F. Swiderski, D. S. Howard, B. A. Grimes, R. M. Rossi, S. J. Szilvassy and C. T. Jordan, Preferential induction of apoptosis for primary human leukemic stem cells, *Proc. Natl. Acad. Sci. U. S. A.*, 2002, **99**, 16220–16225.
30. M. L. Guzman, S. J. Neering, D. Upchurch, B. Grimes, D. S. Howard, D. A. Rizzieri, S. M. Luger and C. T. Jordan, Nuclear factor-kappaB is constitutively activated in primitive human acute myelogenous leukemia cells, *Blood*, 2001, **98**, 2301–2307.
31. C. Jenkins, S. Hewamana, A. Gilkes, S. Neelakantan, P. Crooks, K. Mills, C. Pepper and A. Burnett, Nuclear factor-kappaB as a potential therapeutic target for the novel cytotoxic agent LC-1 in acute myeloid leukaemia, *Br. J. Haematol.*, 2008, **143**, 661–671.
32. B. Baumgartner, M. Weber, M. Quirling, C. Fischer, S. Page, M. Adam, C. Von Schilling, C. Waterhouse, C. Schmid, D. Neumeier and K. Brand, Increased I kappa B kinase activity is associated with activated NF-kappa B in acute myeloid blasts, *Leukemia*, 2002, **16**, 2062–2071.
33. Z. Estrov, S. K. Manna, D. Harris, Q. Van, E. H. Estey, H. M. Kantarjian, M. Talpaz and B. B. Aggarwal, Phenylarsine oxide blocks interleukin-1beta-induced activation of the nuclear transcription factor NF-kappaB, inhibits proliferation, and induces apoptosis of acute myelogenous leukaemia cells, *Blood*, 1999, **94**, 2844–2853.
34. C. E. Bueso-Ramos, F. C. Rocha, S. Shishodia, L. J. Medeiros, H. M. Kantarjian, S. Vadhan-Raj, Z. Estrov, T. L. Smith, M. H. Nguyen and B. B. Aggarwal, Expression of constitutively active nuclear-kappa B RelA transcription factor in blasts of acute myeloid leukemia, *Hum. Pathol.*, 2004, **35**, 246–253.
35. B. Delaval, H. Lelievre and D. Birnbaum, Myeloproliferative disorders: the centrosome connection, *Leukemia*, 2005, **19**, 1739–1744.
36. J. Y. Reuther, G. W. Reuther, D. Cortez, A. M. Pendergast and A. S. Baldwin Jr., A requirement for NF-kappaB activation in Bcr-Abl-mediated transformation, *Genes Dev.*, 1998, **12**, 968–981.
37. M. Y. Hsieh and R. A. Van Etten, IKK-dependent activation of NF-κB contributes to myeloid and lymphoid leukemogenesis by BCR-ABL1, *Blood*, 2014, **123**, 2401–2411.
38. D. Kirchner, J. Duyster, O. Ottmann, R. M. Schmid, L. Bergmann and G. Munzert, Mechanisms of Bcr-Abl-mediated NF-kappaB/Rel activation, *Exp. Hematol.*, 2003, **31**, 504–511.

39. G. Munzert, D. Kirchner, O. Ottmann, L. Bergmann and R. M. Schmid, Constitutive NF-kappab/Rel activation in Philadelphia chromosome positive (Ph+) acute lymphoblastic leukemia (ALL), *Leuk. Lymphoma*, 2004, **45**, 1181–1184.
40. S. Yokota, M. Nakao, S. Horiike, T. Seriu, T. Iwai, H. Kaneko, H. Azuma, T. Oka, T. Takeda, A. Watanabe, A. Kikuta, K. Asami, I. Sekine, T. Matsushita, T. Tsuhciya, J. Mimaya, S. Koizumi, M. Miyake, K. Nishikawa, Y. Takaue, Y. Kawano, A. Iwai, Y. Ishida, K. Matsumoto and T. Fujimoto, Mutational analysis of the N-ras gene in acute lymphoblastic leukemia: a study of 125 Japanese pediatric cases, *Int. J. Hematol.*, 1998, **67**, 379–387.
41. S. Fruehling and R. Longnecker, The immunoreceptor tyrosine-based activation motif of Epstein-Barr virus LMP2A is essential for blocking BCR-mediated signal transduction, *Virology*, 1997, **235**, 241–251.
42. R. G. Caldwell, J. B. Wilson, S. J. Anderson and R. Longnecker, Epstein-Barr virus LMP2A drives B cell development and survival in the absence of normal B cell receptor signals, *Immunity*, 1998, **9**, 405–411.
43. C. Mancao and W. Hammerschmidt, Epstein-Barr virus latent membrane protein 2A is a B-cell receptor mimic and essential for B-cell survival, *Blood*, 2007, **110**, 3715–3721.
44. T. F. Barth, J. I. Martin-Subero, S. Joos, C. K. Menz, C. Hasel, G. Mechtersheimer, R. M. Parwaresch, P. Lichter, R. Siebert and P. Möoller, Gains of 2p involving the REL locus correlate with nuclear c-Rel protein accumulation in neoplastic cells of classical Hodgkin lymphoma, *Blood*, 2003, **101**, 3681–3686.
45. J. I. Martín-Subero, S. Gesk, L. Harder, T. Sonoki, P. W. Tucker, B. Schlegelberger, W. Grote, F. J. Novo, M. J. Calasanz, M. L. Hansmann, M. J. Dyer and R. Siebert, Recurrent involvement of the REL and BCL11A loci in classical Hodgkin lymphoma, *Blood*, 2002, **99**, 1474–1477.
46. S. Joos, C. K. Menz, G. Wrobel, R. Siebert, S. Gesk, S. Ohl, G. Mechtersheimer, L. Trümper, P. Möller, P. Lichter and T. F. Barth, Classical Hodgkin lymphoma is characterized by recurrent copy number gains of the short arm of chromosome 2, *Blood*, 2002, **99**, 1381–1387.
47. F. Emmerich, S. Theurich, M. Hummel, A. Haeffker, M. S. Vry, K. Döhner, K. Bommert, H. Stein and B. Dörken, Inactivating I kappa B epsilon mutations in Hodgkin/Reed-Sternberg cells, *J. Pathol.*, 2003, **201**, 413–420.
48. M. Kato, M. Sanada, I. Kato, Y. Sato, J. Takita, K. Takeuchi, A. Niwa, Y. Chen, K. Nakazaki, J. Nomoto, Y. Asakura, S. Muto, A. Tamura, M. Iio, Y. Akatsuka, Y. Hayashi, H. Mori, T. Igarashi, M. Kurokawa, S. Chiba, S. Mori, Y. Ishikawa, K. Okamoto, K. Tobinai, H. Nakagama, T. Nakahata, T. Yoshino, Y. Kobayashi and S. Ogawa, Frequent inactivation of A20 in B-cell lymphomas, *Nature*, 2009, **459**, 712–716.
49. A. Lake, L. A. Shield, P. Cordano, D. T. Chui, J. Osborne, S. Crae, K. S. Wilson, S. Tosi, S. J. Knight, S. Gesk, R. Siebert, R. T. Hay and

R. F. Jarrett, Mutations of NFKBIA, encoding IkappaB alpha, are a recurrent finding in classical Hodgkin lymphoma but are not a unifying feature of non-EBV-associated cases, *Int. J. Cancer*, 2009, **125**, 1334–1342.

50. J. Houldsworth, S. Mathew, P. H. Rao, K. Dyomina, D. C. Louie, N. Parsa, K. Offit and R. S. Chaganti, REL proto-oncogene is frequently amplified in extranodal diffuse large cell lymphoma, *Blood*, 1996, **87**, 25–29.
51. T. D. Gilmore, C. Cormier, J. Jean-Jacques and M. E. Gapuzan, Malignant transformation of primary chicken spleen cells by human transcription factor c-Rel, *Oncogene*, 2001, **20**, 7098–7103.
52. D. Trecca, L. Guerrini, N. S. Fracchiolla, M. Pomati, L. Baldini, A. T. Maiolo and A. Neri, Identification of a tumor-associated mutant form of the NF-kappaB RelA gene with reduced DNA-binding and transactivating activities, *Oncogene*, 1997, **14**, 791–799.
53. A. Neri, N. S. Fracchiolla, E. Roscetti, S. Garatti, D. Trecca, A. Boletini, L. Perletti, L. Baldini, A. T. Maiolo and E. Berti, Molecular analysis of cutaneous B- and T-cell lymphomas, *Blood*, 1995, **86**, 3160–3172.
54. F. Cerimele, T. Battle, R. Lynch, D. A. Frank, E. Murad, C. Cohen, N. Macaron, J. Sixbey, K. Smith, R. S. Watnick, A. Eliopoulos, B. Shehata and J. L. Arbiser, Reactive oxygen signaling and MAPK activation distinguish Epstein–BarrVirus (EBV)-positive versus EBV-negative Burkitt's lymphoma, *Proc. Natl. Acad. Sci. U. S. A.*, 2005, **102**, 175–179.
55. E. Haralambieva, A. H. Banham, C. Bastard, G. Delsol, P. Gaulard, G. Ott, S. Pileri, J. A. Fletcher and D. Y. Mason, Detection by the fluorescence in situ hybridization technique of MYC translocations in paraffin-embedded lymphoma biopsy samples, *Br. J. Haematol.*, 2003, **121**, 49–56.
56. R. Kuppers, U. Klein, M. L. Hansmann and K. Rajewsky, Cellular origin of human B-cell lymphomas, *N. Engl. J. Med.*, 1999, **341**, 1520–1529.
57. K. J. Savage, S. Monti, J. L. Kutok, G. Cattoretti, D. Neuberg, L. De Leval, P. Kurtin, P. Dal Cin, C. Ladd, F. Feuerhake, R. C. Aguiar, S. Li, G. Salles, F. Berger, W. Jing, G. S. Pinkus, T. Habermann, R. Dalla-Favera, N. L. Harris, J. C. Aster, T. R. Golub and M. A. Shipp, The molecular signature of mediastinal large B-cell lymphoma differs from that of other diffuse large B-cell lymphomas and shares features with classical Hodgkin lymphoma, *Blood*, 2003, **102**, 3871–3879.
58. A. Rosenwald, G. Wright, W. C. Chan, J. M. Connors, E. Campo, R. I. Fisher, R. D. Gascoyne, H. K. Muller-Hermelink, E. B. Smeland, J. M. Giltnane, E. M. Hurt, H. Zhao, L. Averett, L. Yang, W. H. Wilson, E. S. Jaffe, R. Simon, R. D. Klausner, J. Powell, P. L. Duffey, D. L. Longo, T. C. Greiner, D. D. Weisenburger, W. G. Sanger, B. J. Dave, J. C. Lynch, J. Vose, J. O. Armitage, E. Montserrat, A. Lopez-Guillermo, T. M. Grogan, T. P. Miller, M. LeBlanc, G. Ott, S. Kvaloy, J. Delabie, H. Holte, P. Krajci, T. Stokke and L. M. Staudt, The use of molecular profiling to predict survival after chemotherapy for diffuse large-B-cell lymphoma, *N. Engl. J. Med.*, 2002, **346**, 1937–1947.

59. R. E. Davis, K. D. Brown, U. Siebenlist and L. M. Staudt, Constitutive nuclear factor kappaB activity is required for survival of activated B cell-like diffuse large B cell lymphoma cells, *J. Exp. Med.*, 2001, **194**, 1861–1874.
60. F. Feuerhake, J. L. Kutok, S. Monti, W. Chen, A. S. LaCasce, G. Cattoretti, P. Kurtin, G. S. Pinkus, L. de Leval, N. L. Harris, K. J. Savage, D. Neuberg, T. M. Habermann, R. Dalla-Favera, T. R. Golub, J. C. Aster and M. A. Shipp, NFkappaB activity, function, and target-gene signatures in primary mediastinal large B-cell lymphoma and diffuse large B-cell lymphoma subtypes, *Blood*, 2005, **106**, 1392–1399.
61. S. Monti, K. J. Savage, J. L. Kutok, F. Feuerhake, P. Kurtin, M. Mihm, B. Wu, L. Pasqualucci, D. Neuberg, R. C. Aguiar, P. Dal Cin, C. Ladd, G. S. Pinkus, G. Salles, N. L. Harris, R. Dalla-Favera, T. M. Habermann, J. C. Aster, T. R. Golub and M. A. Shipp, Molecular profiling of diffuse large B-cell lymphoma identifies robust subtypes including one characterized by host inflammatory response, *Blood*, 2005, **105**, 1851–1861.
62. L. Tracey, A. Perez-Rosado, M. J. Artiga, F. I. Camacho, A. Rodriguez, N. Martinez, E. Ruiz-Ballesteros, M. Mollejo, B. Martinez, M. Cuadros, J. F. Garcia, M. Lawler and M. A. Piris, Expression of the NF-kappaB targets BCL2 and BIRC5/Survivin characterizes small B-cell and aggressive B-cell lymphomas, respectively, *J. Pathol.*, 2005, **206**, 123–134.
63. J. A. Hernandez, K. J. Land and R. W. McKenna, Leukemias, myeloma and other lymphoreticular neoplasms, *Cancer*, 1995, **75**, 381–394.
64. S. Cuní, P. Pérez-Aciego, G. Pérez-Chacón, J. A. Vargas, A. Sánchez, F. M. Martín-Saavedra, S. Ballester, J. García-Marco, J. Jordá and A. Durántez, A sustained activation of PI3K/NF-κB pathway is critical for the survival of chronic lymphocytic leukaemia B cells, *Leukemia*, 2004, **18**, 1391–1400.
65. R. R. Furman, Z. Asgary, J. O. Mascarenhas, H. C. Liou and E. J. Schattner, Modulation of NF-κB activity and apoptosis in chronic lymphocytic leukemia B cells, *J. Immunol.*, 2000, **164**, 2200–2206.
66. S. Hewamana, S. Alghazal, T. T. Lin, M. Clement, C. Jenkins, M. L. Guzman, C. T. Jordan, S. Neelakantan, P. A. Crooks, A. K. Burnett, G. Pratt, C. Fegan, C. Rowntree, P. Brennan and C. Pepper, The NF-κB subunit Rel A is associated with in vitro survival and clinical disease progression in chronic lymphocytic leukemia and represents a promising therapeutic target, *Blood*, 2008, **111**, 4681–4689.
67. L. Lagneaux, A. Delforge, D. Bron, C. De Bruyn and P. Stryckmans, Chronic lymphocytic leukemic B cells but not normal B cells are rescued from apoptosis by contact with normal bone marrow stromal cells, *Blood*, 1998, **91**, 2387–2396.
68. J. A. Burger, M. Burger and T. J. Kipps, Chronic lymphocytic leukemia B cells express functional CXCR4 chemokine receptors that mediate spontaneous migration beneath bone marrow stromal cells, *Blood*, 1999, **94**, 3658–3667.

69. J. Jurlander, The cellular biology of B-cell chronic lymphocytic leukemia, *Crit. Rev. Oncol. Hematol.*, 1998, **27**, 29–52.
70. P. E. Patten, A. G. Buggins, J. Richards, A. Wotherspoon, J. Salisbury, G. J. Mufti, T. J. Hamblin and S. Devereux, CD38 expression in chronic lymphocytic leukemia is regulated by the tumor microenvironment, *Blood*, 2008, **111**, 5173–5181.
71. S. Lanham, T. Hamblin, D. Oscier, R. Ibbotson, F. Stevenson and G. Packham, Differential signaling via surface IgM is associated with VH gene mutational status and CD38 expression in chronic lymphocytic leukemia, *Blood*, 2003, **101**, 1087–1093.
72. F. K. Stevenson and F. Caligaris-Cappio, Chronic lymphocytic leukemia: revelations from the B-cell receptor, *Blood*, 2004, **103**, 4389–4395.
73. M. Hallek, P. L. Bergsagel and K. C. Anderson, Multiple myeloma: increasing evidence for a multistep transformation process, *Blood*, 1998, **91**, 3–21.
74. H. Ni, M. Ergin, Q. Huang, J. Z. Qin, H. M. Amin, R. L. Martinez, S. Saeed, K. Barton and S. Alkan, Analysis of expression of nuclear factor kappa B (NF-kappa B) in multiple myeloma: downregulation of NF-kappa B induces apoptosis, *Br. J. Haematol.*, 2001, **115**, 279–286.
75. C. Fabre, N. Mimura, K. Bobb, S. Y. Kong, G. Gorgun, D. Cirstea, Y. Hu, J. Minami, H. Ohguchi, J. Zhang, J. Meshulam, R. D. Carrasco, Y. T. Tai, P. G. Richardson, T. Hideshima and K. C. Anderson, Dual inhibition of canonical and noncanonical NF-κB pathways demonstrates significant antitumor activities in multiple myeloma, *Clin. Cancer Res.*, 2012, **18**, 4669–4681.
76. A. C. Bharti, N. Donato and B. B. Aggarwal, Curcumin (diferuloylmethane) inhibits constitutive and IL-6-inducible STAT3 phosphorylation in human multiple myeloma cells, *J. Immunol.*, 2003, **171**, 3863–3871.
77. A. C. Bharti, S. Shishodia, J. M. Reuben, D. Weber, R. Alexanian, S. Raj-Vadhan, Z. Estrov, M. Talpaz and B. B. Aggarwal, Nuclear factor-kappaB and STAT3 are constitutively active in CD138+ cells derived from multiple myeloma patients, and suppression of these transcription factors leads to apoptosis, *Blood*, 2004, **103**, 3175–3184.
78. F. E. Davies, A. M. Dring, C. Li, A. C. Rawstron, M. A. Shammas, S. M. O'Connor, J. A. Fenton, T. Hideshima, D. Chauhan, I. T. Tai, E. Robinson, D. Auclair, K. Rees, D. Gonzalez, A. J. Ashcroft, R. Dasgupta, C. Mitsiades, N. Mitsiades, L. B. Chen, W. H. Wong, N. C. Munshi, G. J. Morgan and K. C. Anderson, Insights into the multistep transformation of MGUS to myeloma using microarray expression analysis, *Blood*, 2003, **102**, 4504–4511.
79. Y. Tu, S. Renner, F. Xu, A. Fleishman, J. Taylor, J. Weisz, R. Vescio, M. Rettig, J. Berenson, S. Krajewski, J. C. Reed and A. Lichtenstein, BCL-X expression in multiple myeloma: possible indicator of chemoresistance, *Cancer Res.*, 1998, **58**, 256–262.

80. A. Arlt, A. Gehrz, S. Muerkoster, J. Vorndamm, M. L. Kruse, U. R. Folsch and H. Schafer, Role of NF-kappaB and Akt/PI3K in the resistance of pancreatic carcinoma cell lines against gemcitabine-induced cell death, *Oncogene*, 2003, **22**, 3243–3251.
81. C. E. Denlinger, B. K. Rundall, M. D. Keller and D. R. Jones, Proteasome inhibition sensitises non-small cell lung cancer to gemcitabine-induced apoptosis, *Ann. Thorac. Surg.*, 2004, **78**, 1207–1214.
82. K. C. Das and C. W. White, Activation of NF-κB by antineoplastic agents, *J. Biol. Chem.*, 1997, **272**, 14914–14920.
83. M. Bentires-Alj, V. Barbu, M. Fillet, A. Chariot, B. Relic, N. Jacobs, J. Gielen, M. P. Merville and V. Bours, NF-kappaB transcription factor induces drug resistance through MDR1 expression in cancer cells, *Oncogene*, 2003, **22**, 90–97.
84. U. Brinkmann and M. Eichelbaum, Polymorphisms in the ABC drug transporter gene MDR1, *Pharmacogenomics J.*, 2001, **1**, 59–64.
85. S. D. Catz and J. L. Johnson, Transcriptional regulation of bcl-2 by nuclear factor kappa B and its significance in prostate cancer, *Oncogene*, 2001, **20**, 7342–7351.
86. J. N. Glasgow, J. Qiu, D. Rassin, M. Grafe, T. Wood and J. R. Perez-Pol, Transcriptional regulation of the BCL-X gene by NF-kappaB is an element of hypoxic responses in the rat brain, *Neurochem. Res.*, 2001, **26**, 647–659.
87. H. Nakano, A. Nakajima, S. Sakon-Komazawa, J. H. Piao, X. Xue and K. Okumura, Reactive oxygen species mediate crosstalk between NF-kappaB and JNK, *Cell Death Differ.*, 2006, **13**, 730–737.
88. S. Kreuz, D. Siegmund, P. Scheurich and H. Wajant, NF-kappaB inducers upregulate cFLIP, a cycloheximide-sensitive inhibitor of death receptor signaling, *Mol. Cell. Biol.*, 2001, **21**, 3964–3973.
89. R. Ravi and A. Bedi, NF-kappaB in cancer - a friend turned foe, *Drug Resist. Update*, 2004, **7**, 53–67.
90. Y. Y. Yu, Q. Li and Z. G. Zhu, NF-kappaB as a molecular target in adjuvant therapy of gastrointestinal carcinomas, *Eur. J. Surg. Oncol.*, 2005, **31**, 386–392.
91. S. Cho, Y. Urata, T. Iida, S. Goto, M. Yamaguchi, K. Sumikawa and T. Kondo, Glutathione downregulates the phosphorylation of I kappa B: autoloop regulation of the NF-kappa B-mediated expression of NF-kappa B subunits by TNF-alpha in mouse vascular endothelial cells, *Biochem. Biophys. Res. Commun.*, 1998, **253**, 104–108.
92. R. Schreck, B. Meier, D. N. Mannel, W. Droge and P. A. Baeuerle, Dithiocarbamates as potent inhibitors of nuclear factor kappa B activation in intact cells, *J. Exp. Med.*, 1992, **175**, 1181–1194.
93. F. D'Acquisto, A. Lalenti, A. Lanaro, R. Di Vaio and R. Carnuccio, Local administration of transcription factor decoy oligonucleotides to nuclear factor-kappaB prevents carrageenin-induced inflammation in rat hind paw, *Gene Ther.*, 2000, **7**, 1731–1737.

94. K. A. Higgins, J. R. Perez, T. A. Coleman, K. Dorshkind, W. A. McComas, U. M. Sarmiento, C. A. Rosen and R. Narayanan, Antisense inhibition of the p65 subunit of NF-kappa B blocks tumorigenicity and causes tumor regression, *Proc. Natl. Acad. Sci. U. S. A.*, 1993, **90**, 9901–9905.
95. C. Frelin, V. Imbert, E. Griessinger, A. C. Peyron, N. Rochet, P. Philip, C. Dageville, A. Sirvent, M. Hummelsberger, E. Berard, M. Dreano, N. Sirvent and J. F. Peyron, Targeting NF-kappaB activation via pharmacologic inhibition of IKK2-induced apoptosis of human acute myeloid leukemia cells, *Blood*, 2005, **105**, 804–811.
96. S. P. Hehner, T. G. Hofmann, W. Dröge and M. L. Schmitz, The anti-inflammatory sesquiterpene lactone parthenolide inhibits NF-kappa B by targeting the I kappa B kinase complex, *J. Immunol.*, 1999, **163**, 5617–5623.
97. T. Hideshima, D. Chauhan, T. Kiziltepe, H. Ikeda, Y. Okawa, K. Podar, N. Raje, A. Protopopov, N. C. Munshi, P. G. Richardson, R. D. Carrasco and K. C. Anderson, Biologic sequelae of I{kappa}B kinase (IKK) inhibition in multiple myeloma: therapeutic implications, *Blood*, 2009, **113**, 5228–5236.
98. J. W. Pierce, R. Schoenleber, G. Jesmok, J. Best, S. A. Moore, T. Collins and M. E. Gerritsen, Novel inhibitors of cytokine induced IkappaBalpha phosphorylation and endothelial cell adhesion molecule expression show anti-inflammatory effects in vivo, *J. Biol. Chem.*, 1997, **272**, 21096–21103.
99. B. M. Pickering, S. de Mel, M. Lee, M. Howell, F. Habens, C. L. Dallman, L. A. Neville, K. N. Potter, J. Mann, D. A. Mann, P. W. Johnson, F. K. Stevenson and G. Packham, Pharmacological inhibitors of NFkappaB accelerate apoptosis in chronic lymphocytic leukaemia cells, *Oncogene*, 2007, **26**, 1166–1177.
100. O. Coux, K. Tanaka and A. L. Goldberg, Structure and functions of the 20S and 26S proteasomes, *Annu. Rev. Biochem.*, 1996, **65**, 801–847.
101. M. Cavo, Proteasome inhibitor bortezomib for the treatment of multiple myeloma, *Leukemia*, 2006, **20**, 1341–1352.
102. P. G. Richardson, T. Hideshima, C. Mitsiades and K. Anderson, Proteasome inhibition in hematologic malignancies, *Ann. Med.*, 2004, **36**, 304–314.
103. S. H. Jin, T. I. Kim, D. S. Han, S. K. Shin and W. H. Kim, Thalidomide suppresses the interleukin 1beta-induced NFkappaB signaling pathway in colon cancer cells, *Ann. N. Y. Acad. Sci.*, 2002, **973**, 414–418.
104. N. Mitsiades, C. Mitsiades, V. Poulaki, D. Chauhan, P. Richardson, T. Hideshima, N. Munshi, S. Treon and K. C. Anderson, Apoptotic signaling induced by immunomodulatory thalidomide analogs in human multiple myeloma cells: therapeutic implications, *Blood*, 2002, **99**, 4525–4530.
105. Y. Takada, A. Bhardwaj, P. Potdar and B. B. Aggarwal, Nonsteroidal anti-inflammatory agents differ in their ability to suppress NF-kappaB activation, inhibition of expression of cyclooxygenase-2 and cyclin D1,

and abrogation of tumor cell proliferation, *Oncogene*, 2004, **23**, 9247–9258.
106. C. Pepper, J. G. Mahdi, A. G. Buggins, S. Hewamana, E. Walsby, E. Mahdi, A. Al-Haza'a, A. J. Mahdi, T. T. Lin, L. Pearce, L. Morgan, I. D. Bowen, P. Brennan and C. Fegan, Two novel aspirin analogues show selective cytotoxicity in primary chronic lymphocytic leukaemia cells that is associated with dual inhibition of Rel A and COX-2, *Cell Prolif.*, 2011, **44**, 380–390.
107. F. Habens, N. Srinivasan, F. Oakley, D. A. Mann, A. Ganesan and G. Packham, Novel sulfasalazine analogues with enhanced NF-κB inhibitory and apoptosis promoting activity, *Apoptosis*, 2005, **10**, 481–491.
108. S. Lentzsch, G. Elliott and G. D. Roodman, SDX-308 and SDX-101, non-steroidal anti-inflammatory drugs, as therapeutic candidates for treating haematologic malignancies including myeloma, *Arch. Pharm. Chem. Life Sci.*, 2007, **340**, 511–516.
109. P. C. Everett, J. A. Meyers, A. Makkinje, M. Rabbi and A. Lerner, Pre-clinical assessment of curcumin as a potential therapy for B-CLL, *Am. J. Hematol.*, 2007, **82**, 23–30.
110. A. K. Ghosh, N. E. Kay, C. R. Secreto and T. D. Shanafelt, Curcumin inhibits prosurvival pathways in chronic lymphocytic leukaemia B cells and may overcome their stromal protection in combination with EGCG, *Clin. Cancer Res.*, 2009, **15**, 1250–1258.
111. M. Holmes-McNary and A. S. Baldwin Jr., Chemopreventive properties of trans-resveratrol are associated with inhibition of activation of the IkappaB kinase, *Cancer Res.*, 2000, **60**, 3477–3483.
112. J. J. Ross, J. T. Arnason and H. C. Birnboim, Low concentrations of the feverfew component parthenolide inhibit in vitro growth of tumor lines in a cytostatic fashion, *Planta Med.*, 1999, **65**, 126–129.
113. C. Wu, F. Chen, J. W. Rushing, X. Wang, H. J. Kim, G. Huang, V. Haley-Zitlin and G. He, Antiproliferative activities of parthenolide and golden feverfew extract against three human cancer cell lines, *J. Med. Food*, 2006, **9**, 55–61.
114. S. Zhang, C. N. Ong and H. M. Shen, Critical roles of intracellular thiols and calcium in parthenolide-induced apoptosis in human colorectal cancer cells, *Cancer Lett.*, 2004, **208**, 143–153.
115. A. J. Garcia-Pineres, M. T. Lindenmeyer and I. Merfort, Role of cysteine residues of p65/NF-kappaB on the inhibition by the sesquiterpene lactone parthenolide and N-ethyl maleimide, and on its transactivating potential, *Life Sci.*, 2004, **75**, 841–856.
116. B. H. Kwok, B. Koh, M. I. Ndubuisi, M. Elofsson and C. M. Crews, The anti-inflammatory natural product parthenolide from the medicinal herb Feverfew directly binds to and inhibits IkappaB kinase, *Chem. Biol.*, 2001, **8**, 759–766.
117. S. Hewamana, T. T. Lin, C. Jenkins, A. K. Burnett, C. T. Jordan, C. Fegan, P. Brennan, C. Rowntree and C. Pepper, The novel nuclear

factor-kappaB inhibitor LC-1 is equipotent in poor prognostic subsets of chronic lymphocytic leukemia and shows strong synergy with fludarabine, *Clin. Cancer Res.*, 2008, **14**, 8102–8111.
118. M. F. Romano, A. Petrella, R. Bisogni, M. C. Turco and S. Venuta, Effect of NF-kappaB/Rel inhibition on spontaneous vs chemotherapy-induced apoptosis in AML and normal cord blood CD34 + cells, *Leukemia*, 2003, **17**, 1190–1192.
119. M. F. Romano, A. Lamberti, R. Bisogni, P. Tassone, D. Pagnini, G. Storti, L. Del Vecchio, M. C. Turco and S. Venuta, Enhancement of cytosine arabinoside-induced apoptosis in human myeloblastic leukemia cells by NF-kappa B/Rel- specific decoy oligodeoxynucleotides, *Gene Ther.*, 2000, **7**, 1234–1237.
120. A. Zhou, S. Scoggin, R. B. Gaynor and N. S. Williams, Identification of NF-kappa B-regulated genes induced by TNFalpha utilizing expression profiling and RNA interference, *Oncogene*, 2003, **22**, 2054–2064.
121. K. M. Rahman, P. J. Jackson, C. H. James, B. P. Basu, J. A. Hartley, M. de la Fuente, A. Schatzlein, M. Robson, R. B. Pedley, C. Pepper, K. R. Fox, P. W. Howard and D. E. Thurston, GC-targeted C8-linked pyrrolobenzodiazepine-biaryl conjugates with femtomolar in vitro cytotoxicity and in vivo antitumor activity in mouse models, *J. Med. Chem.*, 2013, **56**, 2911–2935.
122. L. Li, P. Pongtornpipat, T. Tiutan, S. L. Kendrick, S. Park, D. O. Persky, L. M. Rimsza, S. D. Puvvada and J. H. Schatz, Synergistic induction of apoptosis in high-risk DLBCL by BCL2 inhibition with ABT-199 combined with pharmacologic loss of MCL1, *Leukemia*, 2015, **29**, 1702–1712.
123. R. Pan, V. R. Ruvolo, J. Wei, M. Konopleva, J. C. Reed, M. Pellecchia, M. Andreeff and P. P. Ruvolo, Inhibition of Mcl-1 with the pan-Bcl-2 family inhibitor (-)BI97D6 overcomes ABT-737 resistance in acute myeloid leukemia, *Blood*, 2015, **126**, 363–372.
124. N. 1. Johnson-Farley, J. Veliz, S. Bhagavathi and J. R. Bertino, ABT-199, a BH3 mimetic that specifically targets Bcl-2, enhances the antitumor activity of chemotherapy, bortezomib and JQ1 in "double hit" lymphoma cells, *Leuk. Lymphoma*, 2015, **56**, 2146–2152.
125. I. Kiprianova, J. Remy and N. Milosch, Mohrenz IV, Seifert V, Aigner A, Kögel D. Sorafenib Sensitizes Glioma Cells to the BH3 Mimetic ABT-737 by Targeting MCL1 in a STAT3-Dependent Manner, *Neoplasia*, 2015, **17**, 564–573.
126. L. Wang, D. T. Sloper, S. N. Addo, D. Tian, J. W. Slaton and C. Xing, WL-276, an antagonist against Bcl-2 proteins, overcomes drug resistance and suppresses prostate tumor growth, *Cancer Res.*, 2008, **68**, 4377–4383.
127. Z. Wang, W. Song, A. Aboukameel, M. Mohammad, G. Wang, S. Banerjee, D. Kong, S. Wang, F. H. Sarkar and R. M. Mohammad, TW-37, a small-molecule inhibitor of Bcl-2, inhibits cell growth and invasion in pancreatic cancer, *Int. J. Cancer*, 2008, **123**, 958–966.

128. K. Balakrishnan, W. G. Wierda, M. J. Keating and V. Gandhi, Gossypol, a BH3 mimetic, induces apoptosis in chronic lymphocytic leukemia cells, *Blood*, 008, **112**, 1971–1980.
129. K. Balakrishnan, J. A. Burger, W. G. Wierda and V. Gandhi, AT-101 induces apoptosis in CLL B cells and overcomes stromal cell-mediatedMcl-1 induction and drug resistance, *Blood*, 2009, **113**, 149–153.
130. C. Campas, A. M. Cosialls, M. Barragán, D. Iglesias-Serret, A. F. Santidrián, L. Coll-Mulet, M. de Frias, A. Domingo, G. Pons and J. Gil, Bcl-2 inhibitors induce apoptosis in chronic lymphocytic leukemia cells, *Exp. Hematol.*, 2006, **34**, 1663–1669.

CHAPTER 6

Targeting Signal Transducer and Activator of Transcripion (STAT) 3 with Small Molecules

ANDREW E. SHOUKSMITH AND PATRICK T. GUNNING*

Department of Chemical and Physical Sciences, University of Toronto Mississauga, Mississauga, ON L5L 1C6, Canada
*Email: patrick.gunning@utoronto.ca

6.1 Introduction

6.1.1 STAT Structure

STAT3 is one member of a family of structurally and functionally related transcription factors in the mammalian genome and is encoded by the STAT3 gene.[1] Other known STAT proteins in the mammalian genome include STAT1, STAT2, STAT4, STAT5a, STAT5b and STAT6, all of which possess between 750 and 850 amino acid subunits (90–155 kDa) and are encoded by different genes.[2–4] Despite this, the STAT proteins comprise a conserved structure of six domains: an amino terminus, coiled coil domain (CCD), DNA-binding domain (DBD), linker, sarcoma oncogene (Src)-homology (SH) 2 domain and a C-terminal transactivation (TA) domain (Figure 6.1).[5] The N-terminal domain (residues 1 to approximately 130 in a STAT protein) is important for the formation of STAT tetramers that ultimately bind to specific DNA sites and therefore maximizes transcriptional activation by influencing receptor recognisition.[3,5] The CCD (approximately residues 131–320) is the site at which multiple protein–protein interactions

Drug Discovery Series No. 65
Small-molecule Transcription Factor Inhibitors in Oncology
Edited by Khondaker Miraz Rahman and David E. Thurston

Published by the Royal Society of Chemistry, www.rsc.org

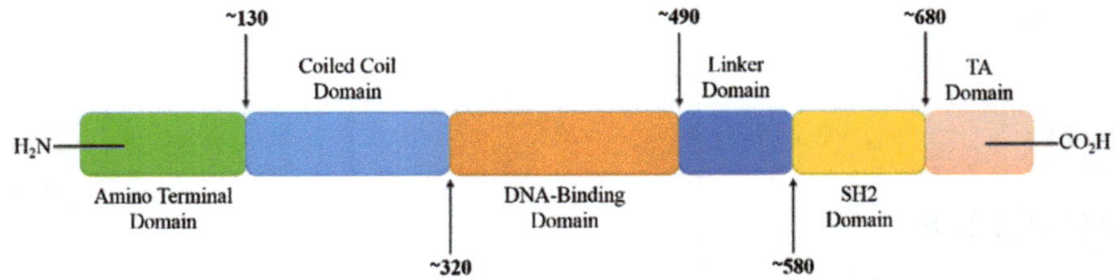

Figure 6.1 General STAT structure, showing the six domains conserved amongst all STAT members and an approximate measure of each domain's length.

(PPIs) are made. Crystal structures obtained for STAT1[4,6] and STAT3[7] have revealed that this region of the protein folds so as to create a predominantly hydrophilic surface that can interact with various proteins, including c-Jun and N-myc-interacting protein (NMI).[3] The DBD (approximately residues 321–490) is similarly structured to the corresponding domains in p53 and NF-κB, containing multiple β-barrels. The number of direct interactions between DNA and the DBD of a particular STAT are relatively modest. Given that STATs bind to DNA as dimers or tetramers, it indicates that cooperative binding is key to achieving the nanomolar dissociation constants (K_D) observed.[3] The linker domain (approximately residues 491–580) connects the DBD with the sarcoma oncogene homology 2 (SH2) domain. Though its exact function is as yet undefined, mutational studies of this domain on STAT1 have indicated that these adversely affect the stability of DNA binding, leading to a rapid off-rate and an inability to activate genes following interferon-gamma (IFN-γ) induction, meaning the linker plays a role in transcriptional regulation.[3,5] The SH2 domain (approximately residues 581–680) is perhaps the most fundamental sequence of the STAT proteins and is the region that is most structurally conserved. It is made up of an anti-parallel β-sheet, flanked by two α-helices, forming a distinct binding pocket. STAT proteins bind to substrates possessing a phosphotyrosine (pY) residue, including other STATs, *via* this SH2 domain, wherein the phosphate ester group interacts with an arginine (R) residue that is universally conserved amongst the STAT family, at the base of the SH2 pocket.[3] The capacity to interact with phosphorylated protein substrates is an essential prerequisite for three STAT signalling events: (1) interaction of the STAT with the cell-membrane-bound cytokine receptor; (2) activation of the STAT *via* interaction with the appropriate kinase; (3) homodimerization and hetero-dimerization with other STATs (see Section 6.1.2 for further detail).[8] Crystal structure analysis of STAT1 has revealed that the amino acids most responsible for defining substrate specificity for the SH2 domain are located at positions +1, +3, +5, +6 and +7, C-terminal from the phosphotyrosine.[3,6] Finally, the TA domain (approximately residues 681 onwards) contains the tyrosine (Y) [and sometimes serine (S)] residue of STATs that is phosphorylated, thereby activating the STAT towards homodimerization and heterodimerization.[9]

6.1.2 STAT Activation and Activity

In healthy cells, all STAT proteins participate in several fundamental cellular processes, including proliferation and differentiation, which gives them a vital role in embryonic development.[10] STATs 2, 4 and 6 are only activated in normal cellular function,[1] playing a vital role in the development of T-cells and in IFN-γ signalling,[2,3] whereas STATs 1, 3, 5a and 5b (the latter two collectively referred to as STAT5 from here on) have been shown to play significant roles in cancer development. More specifically, STAT1 is a known tumour suppressor but STATs 3 and 5 are oncogenes.[1,2,11]

Activation of a STAT protein begins with the coordination of an extracellular cytokine, growth factor or hormone ligand, to a transmembrane tyrosine kinase (TK) receptor (Figure 6.2).[5,10] STATs conduct the signals of a vast library of cytokines, including the interferon family (IFN-α, -β and -γ), the interleukin family (IL-19, -20 and -22), the γC family and the gp-130 family [oncostatin-M (OSM), leukemia inhibitory factor (LIF), cardiotrophin 1 (CT-1) and leptin].[10] As a result of ligand binding, the TK receptor undergoes a conformational shift, which exposes a key Y residue on the cytoplasmic tail of the protein that extends into the cytoplasm. Janus (JAK) kinases, such as JAK1, JAK2, JAK3 and Tyk2, position themselves within sufficiently close proximity to the TK receptor so as to phosphorylate the Y residue on the cytoplasmic tail. This, in turn, creates a docking site on the TK receptor that is recognized by SH2 domain-containing proteins, such as STATs. Trans-phosphorylation occurs at a key Y or S residue within the TA domain of the STAT (Y_{705} or S_{727} for STAT3, Y_{694} for STAT5a and Y_{699} for STAT5b), from the JAK kinase, after which, it dissociates from the receptor.[5] The phosphorylated STAT (pSTAT) undergoes homodimerization or heterodimerization with another pSTAT protein, forming an active dimer, which translocates to the nucleus.[5] The dimers interact with specific gene enhancers, known as gamma-activated site (GAS) elements, and modulate the expression and transcription of genes involved in cell survival called interferon-stimulated genes (ISGs).[10]

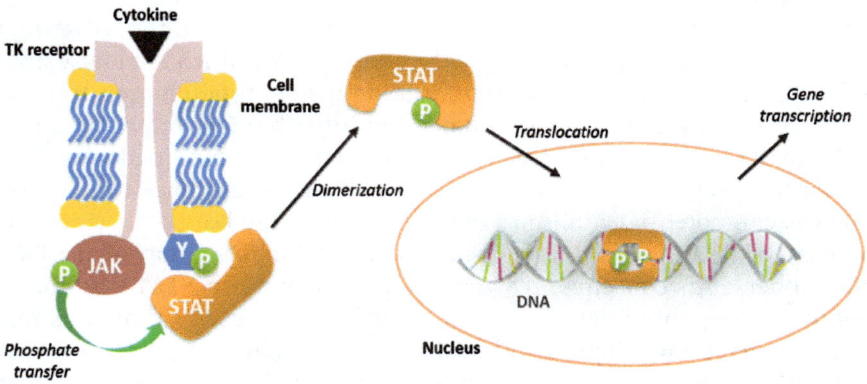

Figure 6.2 General STAT activation mechanism.

Dimerization occurs *via* the reciprocal interaction of the pY residue on one STAT, with the arginine residue located in the SH2 domain of the other. STATs 1, 3, 4 and 5 all form homodimers, but heterodimers have been observed between STATs 1 and 2 and between STATs 1 and 3, depending on the nature and concentration of the activating ligand.[5] Differences in the activation mechanism between the STAT members include the particular JAK that activates the TK receptor towards STAT binding; indeed, JAK kinases have been shown to activate each other by phosphorylating specific Y motifs.[10]

6.1.3 STAT3 as a Target for Cancer Treatment

In multiple human cancers, *e.g.* breast cancer, glioblastoma and blood-borne malignancies, such as leukemia, oncogenic STAT3 is frequently observed, characterized by persistent activation.[12] STAT3 induces the expression of multiple cytokines and growth factors, which, in turn, mediate STAT3 expression, providing a positive feed-forward loop between the tumour cells and the immune cells that keeps STAT3 active to promote tumour propagation. STAT3 impedes both innate and T-cell-mediated anti-tumour immune responses, by antagonizing the activity of nuclear factor (NF)-κB- and STAT1-mediated expression of the T helper 1 (T_H1) cytokines, *e.g.* IL-12 and IFN-γ.[12] In innate immune cells (including macrophages, natural killer cells and neutrophils that provide immediate, non-specific defence against pathogens), STAT3 expression has an immunosuppressive and tumour-promoting influence.[12] This evidence has resulted in the establishment of multiple drug discovery programs for the inhibition of STAT3, as a mode of cancer therapy. Since gene transcription cannot occur without the prior formation of an active STAT homodimer or heterodimer, current efforts to target STAT3 aim to inhibit the protein–protein interaction (PPI) that occurs between two STAT3 monomers. Targeting PPIs in cancer therapy has often been regarded as problematic next to the more established strategy of targeting enzymes, *e.g.* kinases, since the latter often possess readily identifiable active sites, in which potential inhibitors can interact.[13,14] The surfaces of many proteins are, contrastingly, undefined and flat, covering a large surface area ranging from 1150 to over 4000 $Å^2$, which standard small molecules would struggle to bind to, leading many protein targets to be labelled as 'undruggable'.[13,15] Fortunately, within a binding interface, a relatively small number of amino acids are important for a binding interaction, termed 'hot-spots'.[14] Small molecules, with a molecular mass of less than 1000 Da, can potentially inhibit PPIs by binding to the hot-spot residues. In the case of STAT3, the Y_{705} residue is critical for formation of a STAT3 homodimer and, consequently, many reported putative STAT3 inhibitors contain a phosphate ester or other suitable bioisosteres to mimic this residue. This chapter will highlight the advances in this field of cancer research, from 2010 up to 2015 inclusive, although older publications will be referenced as required and patents will be referenced where no journal

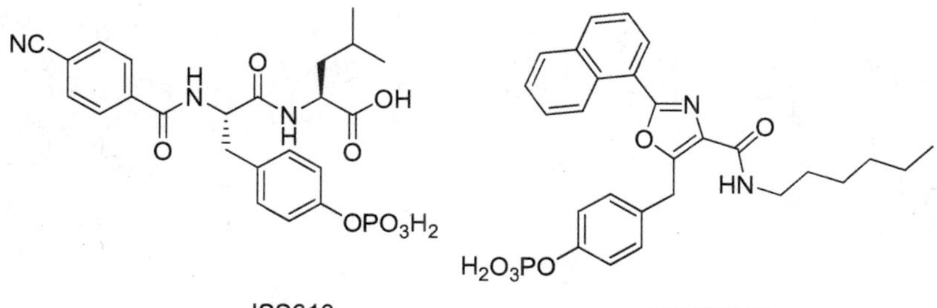

Figure 6.3 Peptidomimetics **1** and **2**.

publication is currently available. The prime focus of this chapter will be on small-molecule-mediated inhibition, but will also describe briefly other types of STAT3 inhibitors, including peptides, oligonucleotides and metal-based complexes.

6.2 Small-molecule STAT3 Inhibitors

6.2.1 Phosphate-ester-containing Inhibitors

The pSTAT3 protein recognizes and binds to phosphorylated STAT monomers, so the majority of first-generation small-molecule STAT3 inhibitors utilized a phosphate ester functionality to achieve potent binding affinity. However, metabolic lability of the phosphate ester, coupled with its poor cell-penetrating properties, have severely curtailed their utility, as seen in the case of the peptidomimetic ISS610 (**1**) (Figure 6.3), where millimolar concentrations were required to elicit an effect in cells.[16,17] To reduce its peptidic character and improve its cell permeability, Hamilton and co-workers incorporated a rigid oxazole or thiazole scaffold to project substituents into the three binding regions of the STAT3 SH2 domain. S3I-M2001 (**2**) became one of the first recorded small-molecule STAT3 inhibitors, disrupting STAT3 dimerization in an electrophoretic mobility shift assay (EMSA) [concentration giving 50% of maximum inhibition (IC_{50}) = 79 μM] with a twofold selectivity over STAT1 (IC_{50} = 159 μM).[18] The mode of action appears to be based on STAT3 dimer–DNA disruption, with promising disruption of the ternary complex observed in a number a cell lines. Suppression of tumour growth was also reported in human breast (MDA-MB-231) tumour-bearing mice when **2** was administered (intravenously 5, 10 and 20 mg kg^{-1} every 2–3 days).

6.2.2 S3I-201

In 2007, Turkson and co-workers reported NSC-74859 (S3I-201, **3**) (Figure 6.4), a non-phosphorylated small molecule which returned an IC_{50} value for the STAT3 homodimer of 86 ± 33 μM using an EMSA.[19] This was in contrast to

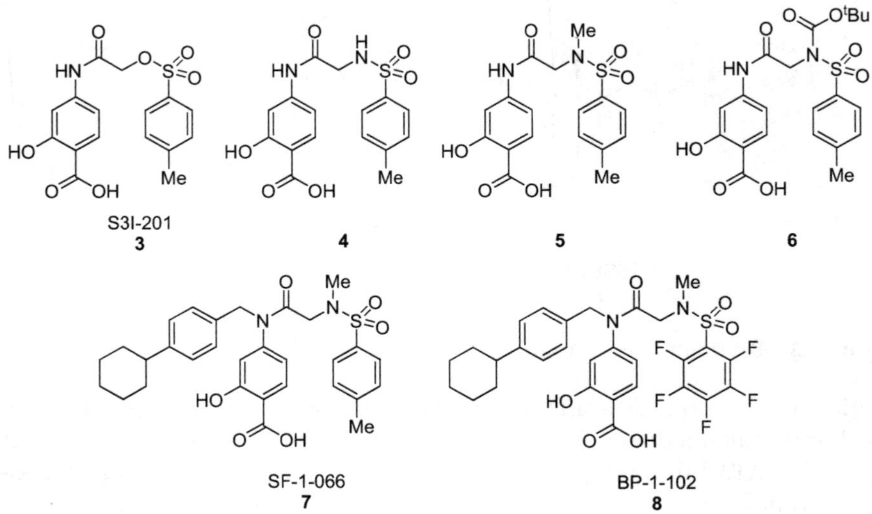

Figure 6.4 4-Aminosalicylic acids **3–8**.

the STAT1–STAT3 heterodimer (160 ± 43 μM), the STAT1 homodimer (over 300 μM) and the STAT5 homodimer (166 ± 17 μM). The *in silico* docking of **3** in the STAT3 SH2 domain indicated that the carboxylic acid group formed hydrogen bonds to S_{611}, S_{613} and R_{609}. The phenolic hydroxyl group made a hydrogen bond with K_{591}, while the tolyl group extended into a partially hydrophobic pocket containing the tetramethylene side chain of K_{592} and the trimethylene side chain of R_{595}, as well as I_{597} and I_{634}.[19] This compound was subsequently used as the basis for other programs to find more drug-like STAT3 inhibitors.

6.2.3 Inhibitors Based on 4-Aminosalicylic Acid

Since compound **3** possessed a tosylate leaving group, it was postulated that **3** may behave as an irreversible STAT3 inhibitor. Gunning and co-workers replaced the in-chain oxygen of the tosylate with an isosteric amino NH (**4**), N-CH_3 (**5**) and N-Boc (**6**) [Boc = $CO_2C(CH_3)_3$] (Figure 6.4). However, these modifications abrogated STAT3 activity relative to **3**.[20] The predicted binding conformation of **3** indicated that attaching substituents on the amide NH would project into another hydrophobic sub-pocket, which contained W_{623}, V_{637}, I_{659} and F_{716}. Using **5** as a common scaffold, various hydrophobic substituents were attached to the amide nitrogen. SF-1-066 (**7**), equipped with a cyclohexylbenzyl group, gave a potency of 35 ± 9 μM by EMSA[20–23] and nuclear extracts from NIH 3T3/v-Src mouse fibroblasts, human Panc-1 pancreatic cancer cells and MDA-MB-231 breast cancer cells showed a dose-dependent reduction in constitutive STAT3 activation upon treatment with **7**.[21] Moreover, surface plasmon resonance (SPR) analysis of **7**, using His-tagged STAT3 (or SH2 domain), returned a K_D of 2.74 μM.[21] Docking studies, using GOLD software, of **7** in the STAT3 SH2 domain indicated a similar binding conformation to that of **3**.[23,24] A structure–activity relationship (SAR) study around the 4-tolyl group resulted in the discovery of BP-1-102 (**8**), with the pentafluorophenyl motif doubling STAT3 potency by EMSA analysis (**8** EMSA IC_{50}: 19.7 ± 5.4 μM) (Figure 6.4).[20,23,25,26] In cell-based studies, **8** displayed between two-fold and six-fold greater activity than **7** in MDA-MB-468 breast cancer cells, DU145 pancreatic cancer cells and JJN3 multiple myeloma (MM) cells.[23] SPR analysis of **8** gave a K_D of 504 nM against STAT3 and, by western blot, **8** effected pSTAT3 knockdown at 20 μM in MDA-MB-468 and JJN3 cells, reducing STAT3 downstream targets, including the cellular homolog of the myelocytomatosis viral oncogene (cMyc), B-cell leukaemia x large (Bcl-xL) and Survivin.[23,25]

When the *N*-methyl group of **7** was replaced with Boc (**9**), GOLD docking placed the compound in a similar conformation to **7**, but with the Boc group directed unfavourably away from the protein surface. This would be predicted to reduce STAT3 activity. However, **9** was similarly active against STAT3 [**9** fluorescence polarization (FP) IC_{50} 15.8 ± 0.2 μM *vs.* **7** FP IC_{50} 31.0 ± 9.4 μM] (Figure 6.5).[20,23,26] The conclusion was that larger *N*-sulfonamide groups caused the molecule to adopt an alternative binding conformation, leading to an extensive SAR around the *N*-sulfonamide substituent.[26] Compounds **10** and **11** were approximately twice as active as the analogous parent compounds (**7** and **8**, respectively) in JJN3 MM cells, demonstrating dose-dependent decreases in cell viability after 72 h in a 3-(4,5-dimethylthiazol-2-yl)-2,5-diphenyltetrazolium bromide (MTT) assay and a reduction in pSTAT3 levels after 6 h by western blot (Figure 6.5).[26]

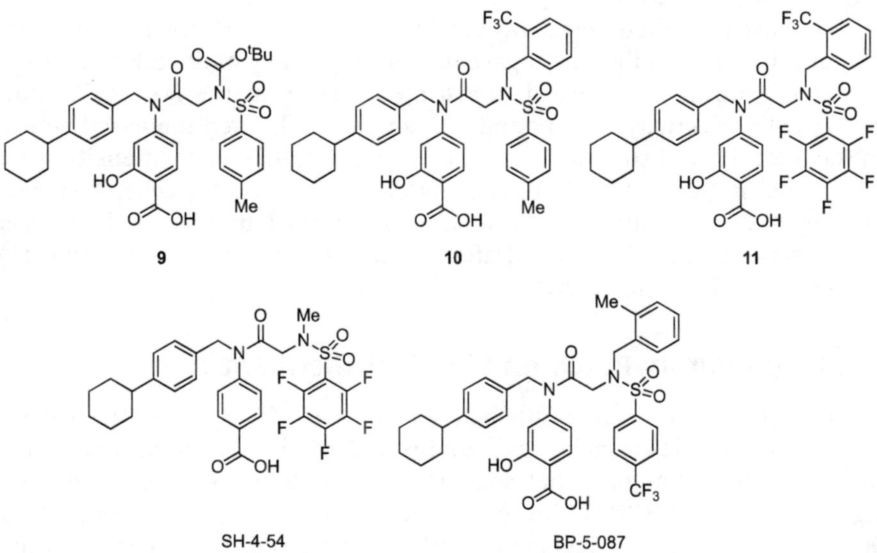

9 10 11

SH-4-54 **12** BP-5-087 **13**

Figure 6.5 4-Aminosalicylic and 4-aminobenzoic acids **9–13**.

However, compound **11** also showed activity in cells that did not harbour pSTAT3, including SKMM2 and MM1.s MM cells. This data would indicate the likelihood of off-target activity and it was of concern that **10** and **11** showed near-identical inhibition of STAT5 and STAT1 phosphorylation as they did inhibition of STAT3 phosphorylation according to phospho-flow cytometry and FP [**10** STAT1 FP K_i 5.8 ± 0.6 μM; STAT3 FP K_i 2.8 ± 4.3 μM; **11** STAT1 FP K_i 10.9 ± 0.8 μM; STAT3 FP K_i 12.8 ± 2.8 μM].[26]

SAR analysis of the salicylic acid motif of **8** provided SH-4-54 (**12**) as one of the first benzoic-acid-based STAT3 inhibitors (Figure 6.5).[27,28] Compound **12** exhibited high potency against multiple brain tumour stem cell (BTSC) lines (25EF, 67EF, 73EF, 84EF and 127EF) with concentration giving 50% of the maximum inhibition of cell proliferation (GI_{50}) values between 66 ± 33.9 nM and 234 ± 73.4 nM. Additionally, it showed minimal toxicity up to 5 μM, in healthy human fetal astrocytes.[28] The high activity of **12** was corroborated using SPR, which returned a K_D of 300 ± 27 nM against STAT3. Western blotting revealed a dose-dependent fall in pY_{705}STAT3 upon treatment with **12**, but not in overall STAT3 levels or on levels of pS_{727}STAT3. Unlike compounds **10** and **11**, compound **12** displayed a four-fold selectivity for STAT3 over STAT1 and negligible activity against kinases, including sarcoma virus oncogene cellular homologue (c-Src), extracellular-signal-regulated kinase (ERK1), Akt, JAK1 and JAK2.[28]

Further SAR studies led to the discovery of BP-5-087 (**13**) (Figure 6.5).[29] Compound **13** was identified using a STAT3-dependant luciferase reporter assay in imatinib-resistant AR230-R chronic myeloid leukemia (CML) cells. The cells were transfected with either STAT3-inducible elements (AR230-R-SIEs) or negative control reporters (AR230-R-NEGs) and compounds were ranked based on their selectivity for SIE cells. While compounds with the pentafluorophenyl substituent gave increased potency in the SIE cells over **7** (positive control), they also demonstrated activity in the NEG cells, indicating poor selectivity. Compound **13**, however, exhibited increased activity in the SIE cells and showed little difference in luminescence intensity in the NEG cells, indicating a STAT3-specific effect. In AR230-R cells, **7** showed little cytotoxic activity, even at 10 μM, in contrast to **13**, which induced apoptosis in these cells at 1 μM after 48 h, as determined using Annexin V staining and flow cytometry.[29]

6.2.4 Inhibitors Based on 5-Aminosalicylic Acid

Sebti and co-workers have published descriptions of STAT3 inhibitors based on a 5-aminosalicylic acid scaffold instead of the 4-aminosalicylic acid motif.[30,31] Interchanging the carboxylic acid and phenol groups did not significantly affect STAT3 binding affinity, with Gunning's compound SF-1-066 (**7**)[20] and Sebti's regioisomeric STAT3 inhibitor (**14**)[31] demonstrating similar activity by FP (**7** K_i 15.5 ± 4.7 μM; **14** K_i 45 ± 12 μM). Sebti and colleagues replaced the sulfonyl glycine backbone with aryl groups that were

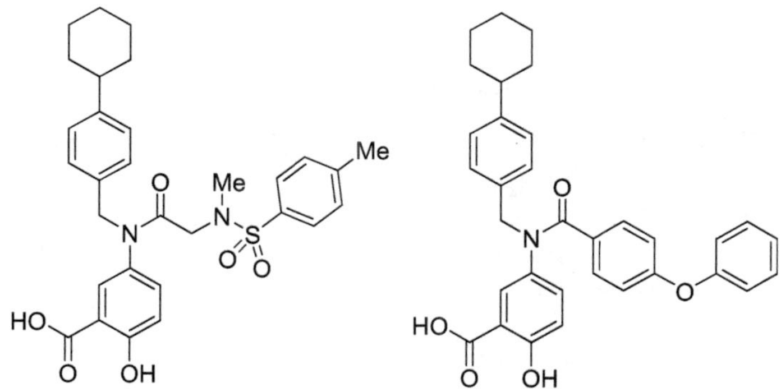

14

S3I-1757
15

Figure 6.6 5-Aminosalicylic acids **14** and **15**.

predicted to increase metabolic stability (Figure 6.6).[30,31] One of these inhibitors, S3I-1757 (**15**), was shown, by *in silico* modelling, to make several interactions with the STAT3 SH2 domain, similar to salicylic acids, **7** and **8**, including to residues R_{609} and K_{591}. Compound **15** displaced the fluorescein-labelled pY gp-130 peptide GpYLPQTV from STAT3 in a FP assay (IC_{50} 15 μM) and inhibition of STAT3–DNA binding was confirmed in MDA-MB-468 nuclear extracts. Treatment of these extracts with a biotin-labelled STAT3–DNA binding probe, followed by **15**, revealed a lower level of STAT3 compared with controls.

6.2.5 Celecoxib

Computational docking using multiple ligand simultaneous docking (MLSD; based on Autodock 4) by Lin and co-workers indicated that the cyclooxygenase 2 (COX2) inhibitor, celecoxib (**16**) could bind to STAT3, with the benzenesulfonamide group occupying the pY_{705} binding pocket and the methylbenzene ring projected into a pocket occupied by L_{706} in the SH2 domain (Figure 6.7).[32–34] In a cell viability assay, Lin and colleagues showed that human rhabdomyosarcoma RD2 cells, pre-treated with **16** (50 or 75 μM) and mixed with 25 ng mL^{-1} IL-6, inhibited IL-6-induced STAT3 phosphorylation by western blot, giving an IC_{50} of 43 μM.[32]

6.2.6 Pyrazole-based Inhibitors

Based on Lin's findings, and using **16** as a starting scaffold, Daniel and co-workers prepared pyrazole-based STAT3 inhibitors designed to bind tightly to the STAT3 SH2 domain, including **17** (Figure 6.7).[35] While these compounds displayed moderate activity, it was postulated that bivalent compounds, able to bind to both SH2 domains of the pSTAT3–pSTAT3 dimer, would prove more effective. On the basis of the results of *in silico* docking

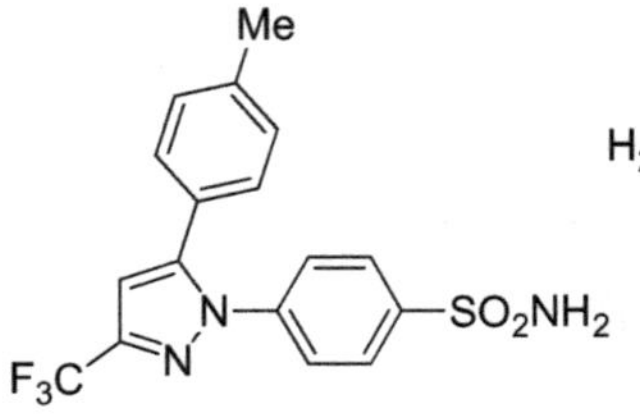

Celecoxib
16

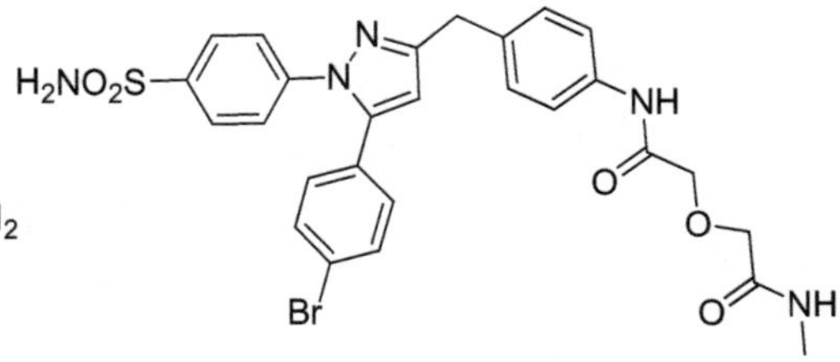

17

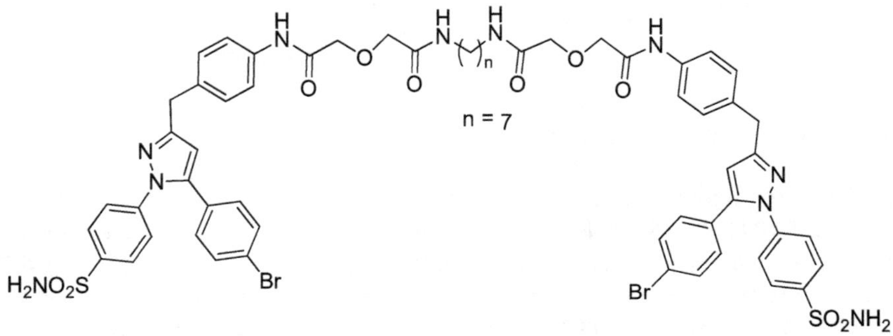

18

Figure 6.7 Pyrazoles **16–18**.

studies it was concluded that the distance between the two pY binding sites in the pSTAT3 dimer was 42 Å, and dimeric pyrazoles bridged by spacers of analogous length were synthesized.[35] Compound **18** (Figure 6.7), a ditopic analogue of **17**, dose-dependently inhibited IL-6-mediated phosphorylation of STAT3 in human glioblastoma dBT114 stem cells and in murine glioblastoma GL261 cells, with concomitant inhibition of downstream targets cyclin D1 and B-cell lymphoma 2 (Bcl-2) after 24 h treatment (1, 10 and 50 μM).[35]

6.2.7 Quinolinecarboxamides

In 2011, Asai and co-workers described a series of quinolinecarboxamides with putative potent STAT3 activity (Figure 6.8).[36–39] In a MTT assay, compounds **19**, **20** and **21** inhibited human lymphoma SCC-3 cell growth, with IC_{50} values of 5.9, 0.9 and 0.3 μM, respectively, indicating inhibition of STAT3 transcriptional activity. Furthermore, in a luciferase reporter assay, **19**, **20** and **21** inhibited 53, 76 and 100% of STAT3 transcriptional activity, respectively, at 100 μM in a HeLa cell line, following treatment with STAT3-activating Oncostatin M. While the preliminary evidence indicates inhibition of the STAT3 pathway, direct STAT3 binding evidence has not been quantified.

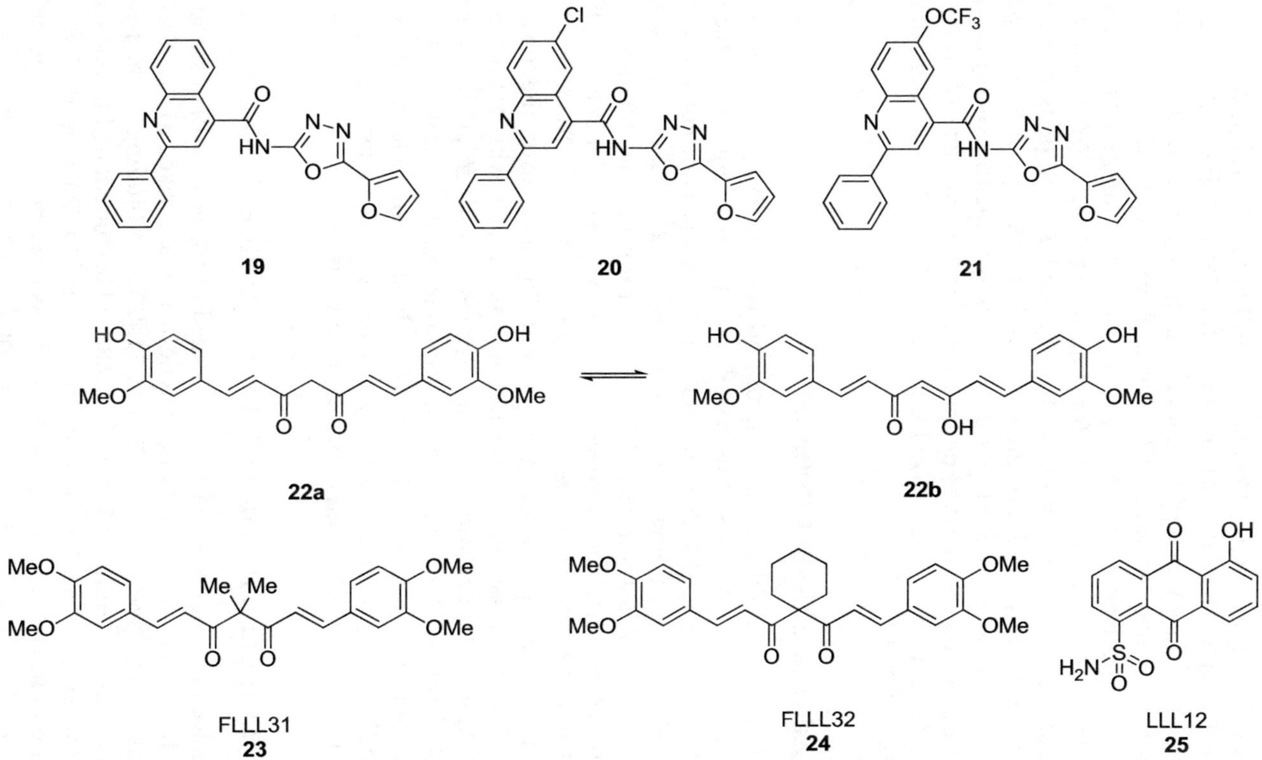

Figure 6.8 Quinolinecarboxamides **19–21**, curcumin (**22**) and analogues **23–24** and LLL12 (**25**).

6.2.8 Curcumin-derived Inhibitors

In 2012, Li and co-workers disclosed the synthesis and JAK2–STAT3 inhibitory activity of curcumin (**22a** and **22b**) and several analogues (Figure 6.8).[40–44] Compounds FLLL31 (**23**) and FLLL32 (**24**) inhibited STAT3 phosphorylation in MDA-MB-231 and MDA-MB-468 cell lines, showing high selectively with no effect against ERK1/2, protein kinase C delta (PKC-δ), mammalian target of rapamycin (mTOR), p70S6K and AKT. SAR studies revealed the significance of stabilizing the diketone tautomer of the curcumin analogues, *via* dialkylation, for improving JAK2–STAT3 selectivity. Compound **24** selectively inhibited IL-6-induced STAT3 phosphorylation without affecting IFN-α-induced phosphorylation of STAT1 or STAT2. While no selectivity was observed for any of these compounds between JAK2 and STAT3, these initial compounds demonstrated that derivatives of curcumin could be used to target the STAT3 protein selectively over the tumour suppressor, STAT1.

6.2.9 Anthraquinone Inhibitors

In addition to curcumin analogues, Li and co-workers have developed a small-molecule STAT3 inhibitor based on an anthraquinone scaffold (Figure 6.8).[42,44] LLL12 (**25**) demonstrated high cytotoxic activity in a diverse array of cell lines, including MDA-MB-231 cells, PANC-1 pancreatic cancer cells and U87 glioblastoma cells (IC_{50} values from 0.16 to 3.09 μM), with demonstrated superior potency over S3I-201 (**3**). Western blot analyses of **25**-treated MDA-MB-231 cell lysates showed decreased pSTAT3 and STAT3-downstream effectors, including Bcl-2 and Survivin, with the onset of caspase-3-induced apoptosis. Xenograft mouse models with MDA-MB-231 and U87 tumours treated with **25** (intraperitoneal, 5 mg kg^{-1}, daily) revealed a two-fold and five-fold reduction in tumour volume, respectively. The results of docking studies indicated that the sulfonamide occupies the pY_{705}-binding sub-pocket of STAT3, forming at least three hydrogen bonds, indicating that the sulfonamide may act as a pY mimetic.

6.2.10 Inhibitors Based on 2-Aminopyridine

Using molecular dynamics (MD) and modelling studies, Williams and co-workers discovered a series of 2-aminopyridines that blocked STAT3–STAT3–DNA binding selectively over STAT1- and STAT5-DNA binding.[45] The results of mutation studies indicated that NSC-368262 (**26**) functioned by alkylating C_{468} in the DBD of STAT3, with an IC_{50} of 10–50 μM (Figure 6.9). The equivalent amino acid in both STATs 1 and 5 is a serine, and compound **26** showed no effect against STAT1, STAT5 or C_{468}S STAT3 mutant binding to DNA by EMSA.[45] In mouse embryonic fibroblasts (MEFs) transfected with a STAT3–yellow fluorescent protein (YFP) fusion protein, **26** (40 μM) completely prevented nuclear accumulation of STAT3–YFP, following

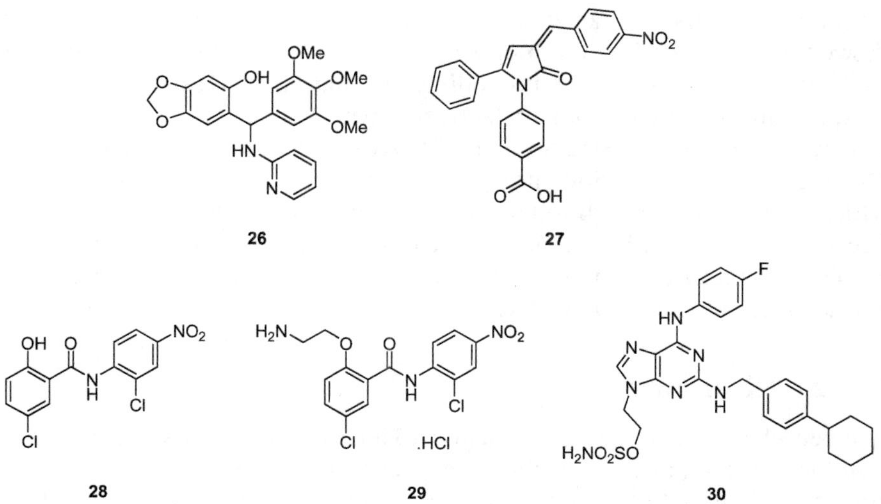

Figure 6.9 2-Aminopyridine **26**, pyrrolidin-2-one **27**, niclosamide (**28**) and its derivative **29** and purine **30**.

stimulation by Oncostatin M. Since the interaction of STAT3 with DNA facilitates nuclear localization, the data support the hypothesis that **26** inhibits STAT3 interaction with DNA.[45] This result was replicated in MDA-MB-231 and MDA-MB-468 cell lines, where **26** (1–20 μM) induced apoptosis. In human prostate cancer LNCaP cells, which do not harbour constitutively active STAT3, there was no effect.[45] In nude mice with subcutaneously induced MDA-MB-468 tumour xenografts, treatment with **26** (intraperitoneal 200 mg kg^{-1}, 5 days per week, 8 weeks total) resulted in significant tumour reduction compared with a control group treated with double distilled H_2O and 2% Tween-80.[45]

6.2.11 Unsaturated Pyrroldin-2-one Inhibitors

By virtually screening approximately 200 000 compounds against the DBD of STAT3, using the crystal structure of STAT3β complexed with DNA [Protein database (PDB) code: 1BG1][7] as a template, Zhang and co-workers identified **27**, which inhibited luciferase reporter expression in MDA-MB-231 cells in a dose-dependent manner (IC_{50} 13.8 ± 0.4 μM) (Figure 6.9).[46] By EMSA, **27** inhibited DNA binding to the STAT3 homodimer, with an IC_{50} of approximately 20 μM, but not to the STAT1 homodimer, even up to 300 μM. Compound **27** also demonstrated a two-fold to four-fold higher selectivity towards various breast and lung cancer cell lines (IC_{50} 3.2–5.4 μM), over the non-cancerous lung fibroblast, IMR90, and mammary epithelial cell line, MCF10A1 (IC_{50} 10–12 μM). In addition, compound **27** (20 μM) was shown to induce apoptosis in MDA-MB-468 cells after 72 h by enzyme-linked immunosorbent assay (ELISA).[46] The mechanism of action of **27** is yet to be determined, but the results of luminescence-based glutathione alkylation assays indicate that, unlike **26**, compound **27** is not an alkylating agent and showed no reduction in glutathione levels in MDA-MB-231 or A549 cells.[46] It also had no effect on STAT3 homodimer formation in H1299 lung cancer cells, in contrast to equivalent cells treated with S3I-201 (**3**), indicating that **27** may not bind to the SH2 domain of STAT3.[46] In a pull-down assay using **27**-conjugated beads, various proteins, in addition to STAT3, were affected. While some of these additional targets could have been pulled down indirectly, as a result of interaction with STAT3, it cannot be ruled out that **27** makes off-target interactions. However, the data supported the hypothesis that **27** binds to the DBD of STAT3.[46]

6.2.12 Niclosamide

A screen of 1500 drug derivatives using a HeLa cell-based STAT3-dependent dual luciferase reporter assay identified niclosamide (**28**) as a potent STAT3 inhibitor (Figure 6.9).[47] Compound **28** was able to dose-dependently reduce STAT3-dependent luciferase reporter activity with an IC_{50} of 0.25 ± 0.07 μM. This effect was also seen in DU145 prostate cancer cells, harbouring activated STAT3, in which **28** (1 μM) suppressed STAT3 Y_{705} phosphorylation within 2 h, without affecting the overall level of the protein. This effect was not observed against isoforms STAT1 or STAT5, nor against upstream activators of STAT3, including JAK1, JAK2 and Src kinases. Compound **28** was shown to induce dose-dependent growth phase 0–1 (G0–G1) cell cycle arrest and apoptosis.[47] Subsequent SAR analysis around the phenoxy hydroxyl group of **28** led to the development of **29**, possessing 3300-fold greater aqueous solubility than **28**, without adversely affecting STAT3 potency (Figure 6.9).[48] Compound **29** was able to suppress MDA-MB-231 breast tumour xenografts in nude mice, both by intraperitoneal (7.5 mg kg^{-1}) and oral (25 mg kg^{-1}) routes, more effectively than **28** at higher doses, which is postulated to be due to its improved solubility resulting in enhanced bioavailability.[48]

6.2.13 Purine-based STAT3 Inhibitors

Gunning and co-workers have previously described the 2-(9*H*-purin-9-yl) acetic acid scaffold as a versatile structure in the creation of potent STAT3 inhibitors.[53] Compound **30** (Figure 6.9) was shown by *in silico* modelling to bind effectively to the STAT3 SH2 domain and induced a potent dose-dependent fall in STAT3-overexpressing XG6 multiple myeloma cell viability, using a MTT cytotoxicity assay, with a GI_{50} of less than 5 μM.[53] Unfortunately, **30** showed cytotoxic activity in SKMM2 cells despite the absence of pSTAT3 in these cells. It was also inactive against JJN3 cells, which overexpress pSTAT3, indicating that the compound was binding to multiple proteins.[54] It was subsequently concluded, using biochemical and biophysical analyses, that the observed activity of **30** in cancer cells was the result of off-target activity not involving STAT3, with evidence to indicate that it interacted with multiple proteins, including members of the JAK family of kinases.[54]

6.2.14 OPB-31121

OPB-31121 is an orally bioavailable STAT3 inhibitor. While its structure has never been disclosed to date, much interest has been garnered around this compound due to its high cellular potency. In a MTT assay, OPB-31121 inhibited Y_{705} and S_{727} phosphorylation of STAT3 in a dose- and time-dependent manner in SNU-484 gastric cancer cells (IC_{50} 5.61 nM).[49] OPB-31121 also displayed less than 10 nM potency in 20 out of 35 hematopoietic cell lines, with KG-1 acute myeloid leukaemia (AML) cells and KU812 chronic myeloid leukaemia (CML) cells recording IC_{50} values below 1 nM.[50] When administered to mice transfected with primary AML cells (orally 200 mg kg^{-1}), cell viability reduced significantly and this effect was selective for the cancer cells over the normal hematopoietic cells.[50] Most recently, virtual docking of OPB-31121 indicated that the compound binds the STAT3 SH2 domain, with at least eight residues in that domain making strong stabilizing interactions with the molecule.[51] Isothermal titration calorimetry (ITC) experiments returned an experimental K_D of 10 nM against the purified STAT3 SH2 domain.[51] As a result, OPB-31121 completed phase I and II clinical trials in humans for treatment of hepatocellular carcinoma (HCC).[52] Unfortunately, the phase II trial was unsuccessful, due to adverse side effects observed at the previously determined therapeutic dose, including changes in blood pressure and pulse rate.[52] Previous research on SNU-484 cells had indicated a promiscuous mechanism of action for OPB-31121. In addition to lowering pSTAT3 levels, OPB-31121 also suppressed JAK2, pJAK2 and gp130, the latter two of which associate in order to phosphorylate STATs. This indicated that OPB-31121 inhibited JAK2 phosphorylation and induced the degradation of JAK2 and gp130.[49]

6.3 Peptide, Oligonucleotide and Organometallic STAT3 Inhibitors

Several publications and patents have described putative STAT3 inhibition with compounds other than small molecules, some representative examples of which are detailed below.

6.3.1 Peptides and Peptidomimetics

McMurray and co-workers have identified peptides with high STAT3 affinity, including the hexapeptide, pYLPQTV, which was derived from residues 904–909 of the gp130 receptor protein. In an EMSA, the hexapeptide recorded an IC_{50} of 150 nM but was predicted to suffer substandard bioavailability.[55,56] The peptide was subsequently modified to enhance metabolic stability and STAT3 affinity, ultimately giving peptidomimetics **31** and **32**, which bound tightly to the STAT3 SH2 domain (IC_{50} 125 nM and 17 nM, respectively) (Figure 6.10).[57] However, *in vivo* the observed activity was significantly lower for both compounds, most probably due to limited cell permeability.

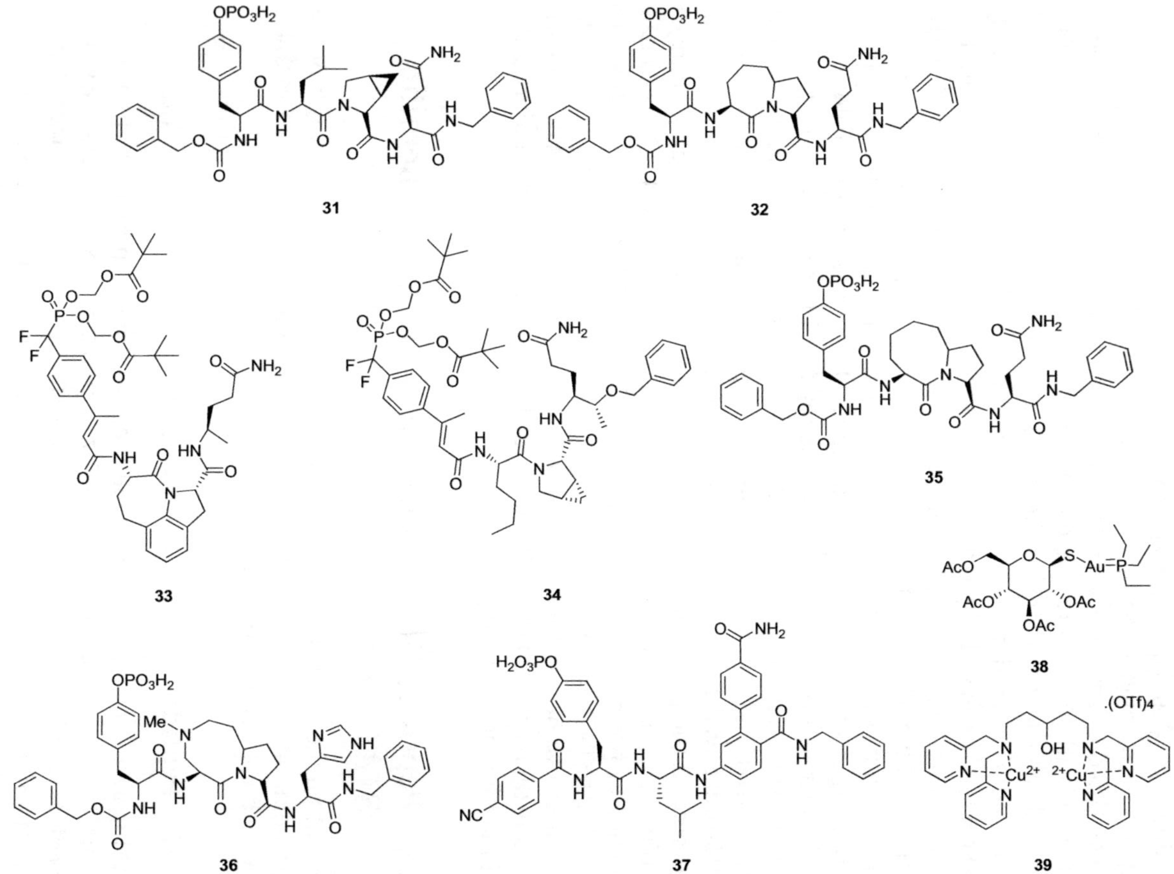

Figure 6.10 Peptides and peptide derivatives **31–37**, auranofin (**38**) and copper complex **39**.

McMurray later patented novel peptidomimetic phosphonate prodrug derivatives of **31**, including **33** and **34**, which targeted the STAT3 SH2 domain with high selectivity over STATs 1 and 5 and increased cellular potency (Figure 6.10).[58] In MDA-MB-468 cells, **33** and **34** (5 μM) completely abrogated STAT3 phosphorylation within 30 min, with the effect sustained for 4 h. Compound **33** (5 μM) also fully suppressed STAT3 phosphorylation in IL-6-stimulated melanoma (MeWo and A375) and ovarian (SKOV3-ip and Hey) cancer cell lines. Finally, mice bearing MDA-MB-468 breast tumour xenografts displayed reduced tumour growth upon administration of **33** (intraperitoneal 170 $mg\,kg^{-1}$, daily for 5 days, with 2 days break over 4 weeks) compared with mice that were given a vehicle [20% Trappsol–phosphate-buffered saline (PBS)]. The compound caused a significant fall in pSTAT3 levels, with no observable organ toxicity.[59]

Using the previously reported gp-130 hexapeptide sequence pYLPQTV as a starting point,[56] Wang and co-workers synthesized several conformationally restricted peptidomimetics, including **35**.[60] This compound demonstrated a K_i of 17 nM in a STAT3-competitive FP assay, making it 20-fold more active than the hexapeptide, but unfortunately, it showed no activity up to 100 μM in MDA-MB-231 and MDA-MB-468 cell lines.[60] SAR analysis revealed that the glutamine side chain of **35** was essential for STAT3 activity in the FP assay, with reduction of the amide or replacement with simple alkyl chains or a histidine ring severely abrogating STAT3 potency between 10-fold and 500-fold.[60] However, the benzyl carbamate of **35** was found to be more amenable to substitution. The replacement of the benzyloxy group with a long alkyl chain ($C_{15}H_{31}$) gave a compound of near-identical potency to **35** but with greatly improved cellular activity. Replacement of the glutamine side chain with that of a histidine reduced FP *in vitro* activity 25-fold relative to **35**, but analogues with a histidine side chain retained their cellular activity, indicating that the primary amide of **35** was detrimental to cell membrane penetration.[60] Finally, to reduce the overall negative charge of the compound arising from the phosphate group, one of the carbons in the eight-membered ring of **35** was replaced with a methylamino group. Compound **36** was five-fold to ten-fold more active than the derivative bearing no methylamino group in MDA-MB-231 and MDA-MB-468 cells, giving an IC_{50} of 11.2 and 3.6 μM, respectively.[60] Compound **36** induced robust cleavage of caspase-3 and poly ADP ribose polymerase (PARP), two apoptosis biomarkers, at 5–10 μM after 2 days, in MDA-MB-468 cells, and reduced levels of pSTAT3 and its downstream targets Bcl-xL and cyclin D1, without affecting overall STAT3 levels.[60] At time of publication, no animal-based data has been disclosed for these compounds.

Gunning and co-workers have also described peptidomimetic inhibitors, formed by the amalgamation of the known STAT3-inhibitory peptidomimetic ISS610 (**1**) with parts of the STAT3-binding gp-130 sequence pYLPQTV.[17] Compound **37** (Figure 6.10) displayed effective STAT3 binding by FP, with an IC_{50} of 5 μM, and modelling data indicated that the biphenylcarboxamide motif was mimicking the binding interaction made by the Pro–Gln dipeptide in Wang's compound **35**.[17] However, while **37** showed no activity against STAT5, it was equipotent against the tumour suppressor STAT1 by FP (IC_{50} = 6.5 μM). It is for this reason, perhaps, that **37** was considerably less active (IC_{50} > 50 μM) against STAT3 DNA binding in NIH3T3/vSrc cells, as determined by EMSA analysis, due to the presence of other protein targets, including STAT1. Compound **37** inhibited approximately 80% of STAT3 in NIH3T3/vSrc cells after 6 h at 100 μM, but after 24 h, STAT3 activity had partially recovered.[17] By SPR using His-tagged STAT3, **37** exhibited high binding affinity to the SH2 domain (K_D = 900 nM). Replacing the phosphate ester with a phenolic hydroxyl group resulted in abrogation of binding affinity (K_D = 12 μM), though interestingly, this derivative was equipotent against STAT3–DNA binding by EMSA in NIH3T3/vSrc cells to the corresponding phosphate ester.[17] Unsurprisingly, this indicates the phosphate ester gets rapidly metabolized to the phenol in cells.

6.3.2 Oligonucleotides

Swayze and co-workers have modulated STAT3 expression using chemically modified nucleotide subunits called gapmers.[61] Several hundred gapmers were screened for the antisense inhibition of STAT3 in human umbilical vein endothelial cells (HUVECs), with three displaying over 90% STAT3 inhibition: ISIS No. 481464, ISIS No. 481549 and ISIS No. 455291. These gapmers possessed the mRNA-binding sequences CTATTTGGATGTCAGC (16 mer), GAAATTCATTCTTCCA (16 mer) and CAGCAGATCAAGTCCAGGGA (20 mer), respectively (structures not disclosed). These compounds were found to be cell-permeable in U251-MG (human glioblastoma astrocytoma), MDA-MB-231, A431 (epidermoid carcinoma) and NCI H460 (lung cancer) cells, where they exhibited low micromolar activity and a half-life longer than 12 days. In MDA-MB-231 tumour xenografts in BALB/c nude mice, treatment with these oligonucleotides (intraperitoneal 25 mg kg^{-1}, twice a week for three weeks), resulted in a reduced rate of tumour growth, however the tumour ultimately continued to advance, showing that these compounds could not provide complete inhibition of the cancer.[61]

6.3.3 Organometallic Complexes

Priebe and co-workers published several gold-complexed thiosaccharides, which abolished STAT3 phosphorylation in a variety of brain cancers (Figure 6.10).[62] Auranofin (**38**) returned GI_{50} values of 1.5, 1.6 and 3.8 μM in BT-58, 58-10 and U-87 cell lines, respectively, and effected a dose-dependent fall in pSTAT3 levels in IFNα- and IL-6-stimulated BT-58 cells. In IFNα- and IL-6-induced HH cutaneous T-cell lymphoma (CTCL) cells, **38** (1 μM) effected ≥50% decrease in cell viability under normoxic and hypoxic conditions, with a 30% decrease observed under normal oxygen perfusion.

Gunning and co-workers have also described copper(II)-containing complexes with novel bis-dipicolylamine (BDPA) frameworks, such as **39** (Figure 6.10).[63,64] Using titration experiments at pH 7.4 (5 mM substrate,

0.25 mM ligand), the complexes gave moderate binding affinities for the STAT3 native peptide sequence (Ac-pYLKTK-NH_2) of 8.2–43 μM.[63] By EMSA analysis, the complexes induced significant STAT3–STAT3 dimerization disruption and the suppression of STAT3–DNA binding in NIH3T3/v-Src cells (IC_{50} 41–74 μM).[63] The unchelated derivatives showed no effect in these cells at up to 300 μM. Suppression of STAT3 dimers was also observed upon treatment of DU-145 and MDA-MB-468 cells with these copper complexes but not in healthy NIH3T3 cells.[63]

6.4 Conclusions

STAT3 knockout studies have demonstrated that a STAT3 therapeutic would be of significant importance in cancer research. There have now been several clinical trials of molecules that target STAT3 activity in human disease. While to-date, a successful STAT3-targeting drug has not entered clinical use; many nanomolar-potency binders have now been identified that demonstrate selectivity for STAT3 amongst the other STAT isoforms, as well as against other upstream kinases. This indicates that STAT3 protein is a tractable drug target and that its binding domains, particularly the SH2 domain, have pockets which can bind a small molecule. Moreover, it seems likely that if the pharmacokinetic properties of such an inhibitor can be suitably optimized, a STAT3-targeting drug will reach clinical use.

References

1. X. Wang, P. J. Crowe, D. Goldstein and J. Yang, *Int. J. Oncol.*, 2012, **41**, 1181.
2. S. Y. Lai and F. M. Johnson, *Drug Resist. Updates*, 2010, **13**, 67.
3. T. Kisseleva, S. Bhattacharya, J. Braunstein and C. W. Schindler, *Gene*, 2002, **285**, 1.
4. X. Mao, Z. Ren, G. N. Parker, H. Sondermann, M. A. Pastorello, W. Wang, J. S. McMurray, B. Demeler, J. E. Darnell and X. Chen, *Mol. Cell*, 2005, **17**, 761.
5. D. E. Levy and J. E. Darnell, *Nat. Rev. Mol. Cell Biol.*, 2002, **3**, 651.
6. X. Chen, U. Vinkemeier, Y. Zhao, D. Jeruzalmi, J. E. Darnell and J. Kuriyan, *Cell*, 1998, **93**, 827.
7. S. Becker, B. Groner and C. W. Muller, *Nature*, 1998, **394**, 145.
8. S. Gupta, H. Yan, L. H. Wong, S. Ralph, J. Krolewski and C. Schindler, *EMBO J.*, 1996, **15**, 1075.
9. B. B. Aggarwal, A. B. Kunnumakkara, K. B. Harikumar, S. R. Gupta, S. T. Tharakan, C. Koca, S. Dey and B. Sung, *Ann. N. Y. Acad. Sci.*, 2009, **1171**, 59.
10. D. Domańska and E. Brzeziańska, *Curr. Signal Transduction Ther.*, 2012, **7**, 187.
11. K. M. Quesnelle, A. L. Boehm and J. R. Grandis, *J. Cell. Biochem.*, 2007, **102**, 311.

12. H. Yu, D. Pardoll and R. Jove, *Nat. Rev. Cancer*, 2009, **9**, 798.
13. J. A. Wells and C. L. McClendon, *Nature*, 2007, **450**, 1001.
14. A. W. White, A. D. Westwell and G. Brahemi, *Expert Rev. Mol. Med.*, 2008, **10**, 1.
15. L. L. Conte, C. Chothia and J. Janin, *J. Mol. Biol.*, 1999, **285**, 2177.
16. K. A. Z. Siddiquee, P. T. Gunning, M. Glenn, W. P. Katt, S. Zhang, C. Schroeck, S. M. Sebti, R. Jove, A. D. Hamilton and J. Turkson, *ACS Chem. Biol.*, 2007, **2**, 787.
17. V. M. Shahani, P. Yue, S. Fletcher, S. Sharmeen, M. A. Sukhai, D. P. Luu, X. Zhang, H. Sun, W. Zhao, A. D. Schimmer, J. Turkson and P. T. Gunning, *Bioorg. Med. Chem.*, 2011, **19**, 1823.
18. J. Turkson, A. D. Hamilton, R. Jove and S. M. Sebti, STAT3 Inhibitor Having Anti-Cancer Activity and Associated Methods. EP 2120958 B1, 2013.
19. K. Siddiquee, S. Zhang, W. C. Guida, M. A. Blaskovich, B. Greedy, H. R. Lawrence, M. L. R. Yip, R. Jove, M. M. McLaughlin, N. J. Lawrence, S. M. Sebti and J. Turkson, *Proc. Natl. Acad. Sci.*, 2007, **104**, 7391.
20. S. Fletcher, J. Singh, X. Zhang, P. Yue, B. D. G. Page, S. Sharmeen, V. M. Shahani, W. Zhao, A. D. Schimmer, J. Turkson and P. T. Gunning, *ChemBioChem*, 2009, **10**, 1959.
21. X. Zhang, P. Yue, S. Fletcher, W. Zhao, P. T. Gunning and J. Turkson, *Biochem. Pharmacol.*, 2010, **79**, 1398.
22. S. Fletcher, B. D. G. Page, X. Zhang, P. Yue, Z. H. Li, S. Sharmeen, J. Singh, W. Zhao, A. D. Schimmer, S. Trudel, J. Turkson and P. T. Gunning, *ChemMedChem*, 2011, **6**, 1459.
23. B. D. G. Page, S. Fletcher, P. Yue, Z. H. Li, X. Zhang, S. Sharmeen, A. Datti, J. L. Wrana, S. Trudel, A. D. Schimmer, J. Turkson and P. T. Gunning, *Bioorg. Med. Chem. Lett.*, 2011, **21**, 5605.
24. D. Resetca, S. Haftchenary, P. T. Gunning and D. J. Wilson, *J. Biol. Chem.*, 2014, **289**, 32538.
25. X. Zhang, P. Yue, B. D. G. Page, T. Li, W. Zhao, A. T. Namanja, D. Paladino, J. Zhao, Y. Chen, P. T. Gunning and J. Turkson, *Proc. Natl. Acad. Sci.*, 2012, **109**, 9623.
26. B. D. G. Page, D. C. Croucher, Z. H. Li, S. Haftchenary, V. H. Jimenez-Zepeda, J. Atkinson, P. A. Spagnuolo, Y. L. Wong, R. Colaguori, A. M. Lewis, A. D. Schimmer, S. Trudel and P. T. Gunning, *J. Med. Chem.*, 2013, **56**, 7190.
27. P. Yue, S. Haftchenary, P. T. Gunning and J. Turkson, *Cancer Res.*, 2013, **73**(8 suppl.), 3253.
28. S. Haftchenary, H. A. Luchman, A. O. Jouk, A. J. Veloso, B. D. G. Page, X. R. Cheng, S. S. Dawson, N. Grinshtein, V. M. Shahani, K. Kerman, D. R. Kaplan, C. Griffin, A. M. Aman, R. Al-awar, S. Weiss and P. T. Gunning, *ACS Med. Chem. Lett.*, 2013, **4**, 1102.
29. A. M. Eiring, B. D. G. Page, I. L. Kraft, C. C. Mason, N. A. Vellore, D. Resetca, M. S. Zabriskie, T. Y. Zhang, J. S. Khorashad, A. J. Engar, K. R. Reynolds, D. J. Anderson, A. Senina, A. D. Pomicter, C. C. Arpin,

S. Ahmad, W. L. Heaton, S. K. Tantravahi, R. Colaguori, R. Moriggl, D. J. Wilson, R. Baron, T. O'Hare, P. T. Gunning and M. W. Deininger, *Leukemia*, 2015, **29**, 586.
30. X. Zhang, Y. Sun, R. Pireddu, H. Yang, M. K. Urlam, H. R. Lawrence, W. C. Guida, N. J. Lawrence and S. M. Sebti, *Cancer Res.*, 2013, **73**, 1922.
31. M. K. Urlam, R. Pireddu, Y. Ge, X. Zhang, Y. Sun, H. R. Lawrence, W. C. Guida, S. M. Sebti and N. J. Lawrence, *Med. Chem. Commun.*, 2013, **4**, 932.
32. S. Reed, H. Li, C. Li and J. Lin, *Biochem. Biophys. Res. Commun.*, 2011, **407**, 450.
33. Y. Liu, A. Liu, H. Li, C. Li and J. Lin, *Cancer Prev. Res.*, 2011, **4**, 1296.
34. H. Li, A. Liu, Z. Zhao, Y. Xu, J. Lin, D. Jou and C. Li, *J. Med. Chem.*, 2011, **54**, 5592.
35. D. J. Daniel, I. F. Parney and T. E. Peterson, Pyrazole Derivatives as Inhibitors of STAT3, WO 2013187965 A1, 2013.
36. K. Matsuno, Y. Masuda, Y. Uehara, H. Sato, A. Muroya, O. Takahashi, T. Yokotagawa, T. Furuya, T. Okawara, M. Otsuka, N. Ogo, T. Ashizawa, C. Oshita, S. Tai, H. Ishii, Y. Akiyama and A. Asai, *ACS Med. Chem. Lett.*, 2010, **1**, 371.
37. T. Ashizawa, H. Miyata, H. Ishii, C. Oshita, K. Matsuno, Y. Masuda, T. Furuya, T. Okawara, M. Otsuka, N. Ogo, A. Asai and Y. Akiyama, *Int. J. Oncol.*, 2011, **38**, 1245.
38. T. Ashizawa, H. Miyata, A. Iizuka, M. Komiyama, C. Oshita, A. Kume, M. Nogami, M. Yagoto, I. Ito, T. Oishi, R. Watanabe, K. Mitsuya, K. Matsuno, T. Furuya, T. Okawara, M. Otsuka, N. Ogo, A. Asai, Y. Nakasu, K. Yamaguchi and Y. Akiyama, *Int. J. Oncol.*, 2013, **43**, 219.
39. T. Ashizawa, Y. Akiyama, H. Miyata, A. Iizuka, M. Komiyama, A. Kume, M. Omiya, T. Sugino, A. Asai, N. Hayashi, K. Mitsuya, Y. Nakasu and K. Yamaguchi, *Int. J. Oncol.*, 2014, **45**, 411.
40. M. A. Bill, J. R. Fuchs, C. Li, J. Yui and C. Bakan, *Mol. Cancer*, 2010, **9**, 165.
41. S. L. Fossey, M. D. Bear, J. Lin, C. Li, E. B. Schwartz, P. Li, J. R. Fuchs, J. Fenger, W. C. Kisseberth and C. A. London, *BMC Cancer*, 2011, **11**, 112.
42. C. Wei, S. Ball, L. Lin, A. Liu, J. R. Fuchs, P. Li, C. Li and J. Lin, *Int. J. Oncol.*, 2011, **38**, 279.
43. M. A. Bill, C. Nicholas, T. A. Mace, J. P. Etter, C. Li, E. B. Schwartz, J. R. Fuchs, G. S. Young, L. Lin, J. Lin, L. He, M. Phelps, P. Li and G. B. Lesinski, *PLoS One*, 2012, **7**, e40724.
44. G. Onimoe, A. Liu, L. Lin, C. Wei, E. B. Schwartz, D. Bhasin, C. Li, J. R. Fuchs, P. Li, P. Houghton, A. Termuhlen, T. Gross and J. Lin, *Invest. New Drugs*, 2012, **30**, 916.
45. R. Buettner, R. Corzano, R. Rashid, J. Lin, M. Senthil, M. Hedvat, A. Schroeder, A. Mao, A. Herrmann, J. Yim, H. Li, Y. Yuan, K. Yakushijin, F. Yakushijin, N. Vaidehi, R. Moore, G. Gugiu, T. D. Lee, R. Yip, Y. Chen, R. Jove, D. Horne and J. C. Williams, *ACS Chem. Biol.*, 2011, **6**, 432.
46. W. Huang, Z. Dong, F. Wang, H. Peng, J. Liu and J. Zhang, *ACS Chem. Biol.*, 2014, **9**, 1188.

47. X. Ren, L. Duan, Q. He, Z. Zhang, Y. Zhou, D. Wu, J. Pan, D. Pei and K. Ding, *ACS Med. Chem. Lett.*, 2010, **1**, 454.
48. H. Chen, Z. Yang, C. Ding, L. Chu, Y. Zhang, K. Terry, H. Liu, Q. Shen and J. Zhou, *ACS Med. Chem. Lett.*, 2013, **4**, 180.
49. M. Kim, H. Nam, H. Kim, S. Han, S. Im, T. Kim, D. Oh and Y. Bang, *Cancer Lett.*, 2013, **335**, 145.
50. F. Hayakawa, K. Sugimoto, Y. Harada, N. Hashimoto, N. Ohi, S. Kurahashi and T. Naoe, *Blood Cancer J.*, 2013, **3**, e166.
51. L. Brambilla, D. Genini, E. Laurini, J. Merulla, L. Perez, M. Fermeglia, G. M. Carbone, S. Pricl and C. V. Catapano, *Mol. Oncol.*, 2015, **9**, 1194.
52. Phase I/II Study of OPB-31121 in Patients with Progressive Hepatocellular Carcinoma, (accessed 24 July).
53. V. M. Shahani, P. Yue, S. Haftchenary, W. Zhao, J. L. Lukkarila, X. Zhang, D. Ball, C. Nona, P. T. Gunning and J. Turkson, *ACS Med. Chem. Lett.*, 2010, **2**, 79.
54. V. M. Shahani, D. P. Ball, A. V. Ramos, Z. Li, P. A. Spagnuolo, S. Haftchenary, A. D. Schimmer, S. Trudel and P. T. Gunning, *Bioorg. Med. Chem.*, 2013, **21**, 5618.
55. Z. Y. Ren, L. A. Cabell, T. S. Schaefer and J. S. McMurray, *Bioorg. Med. Chem. Lett.*, 2003, **13**, 633.
56. C. Gomez, L. Bai, J. Zhang, Z. Nikolovska-Coleska, J. Chen, H. Yi and S. Wang, *Bioorg. Med. Chem. Lett.*, 2009, **19**, 1733.
57. J. S. McMurray, Z. Y. Ren, D. R. Coleman I.V., P. K. Mandal, X. Chen and W. Liao, Inhibitors of Signal Transducer and Activator of Transcription 3, US 20070010428 A1, 2007.
58. P. K. Mandal, F. Q. Gao, Z. Lu, Z. Y. Ren, R. Ramesh, J. S. Birtwistle, K. K. Kaluarachchi, X. M. Chen, R. C. Bast, W. S. Liao and J. S. McMurray, *J. Med. Chem.*, 2011, **54**, 3549.
59. J. S. McMurray, P. K. Mandal, W. S. Liao, J. Klostergaard and F. M. Robertson, *JAK-STAT*, 2012, **1**, 263.
60. J. Chen, L. Bai, D. Bernard, Z. Nikolovska-Coleska, C. Gomez, J. Zhang, H. Yi and S. Wang, *ACS Med. Chem. Lett.*, 2010, **1**, 85.
61. S. M. Freier, Y. Kim, R. A. Macleod and E. E. Swayze, Modulation of Signal Transducer and Activator of Transcription 3 (Stat3) Expression, WO 2012135736 A2, 2012.
62. C. Conrad, I. Fokt, T. Kato, T. Madden, W. Priebe and S. Skora, Auranofin and Auranofin Analogs Useful to Treat Proliferative Disease and Disorders, WO 2012142615 A2, 2012.
63. J. A. Drewry, S. Fletcher, P. Yue, D. Marushchak, W. Zhao, S. Sharmeen, X. Zhang, A. D. Schimmer, C. Gradinaru, J. Turkson and P. T. Gunning, *Chem. Commun.*, 2010, **46**, 892.
64. J. A. Drewry, S. Fletcher, H. Hassan and P. T. Gunning, *Org. Biomol. Chem.*, 2009, **7**, 5074.

CHAPTER 7

Targeting Promoter G-Quadruplexes for Transcriptional Control

RHIANNA K. MORGAN[a] AND TRACY A. BROOKS*[a,b]

[a] Department of BioMolecular Sciences, Division of Pharmacology, School of Pharmacy, University of Mississippi, University, MS 38677, USA;
[b] Department of Pharmaceutical Sciences, School of Pharmacy, Binghamton University, Binghamton, NY 13902, USA
*Email: tbrooks@binghamton.edu

7.1 G-Quadruplex Formation within the Genome

Before the structure of DNA was predicted by Watson and Crick, a German chemist named Ivar Bang described gel formation in aqueous solution at high concentrations of guanosine residues, guanylic acid, or its derivatives in 1910.[1] This unique behavior sparked the initial interest in guanine bases. Though guanine gel formation was a common frustration in many laboratories, it was fifty years later that the first fiber X-ray diffraction data looked into the phenomenon of the guanine tetrameric arrangement referred to as a G-quartet.[2] A G-quartet, now referred to as a tetrad, forms in guanine-rich sequences of DNA or RNA. Its formation requires two or more guanine repeats within a sequence that each consist of at least four residues. These residues bond with each other in a plane *via* Hoogsteen base-pairing with each guanine acting as both donor and acceptor of two hydrogen bonds. With a pocket at its center, featuring electronegative carbonyl oxygens, G-quartets have increased stability with a monovalent cation like K^+.[3]

Drug Discovery Series No. 65
Small-molecule Transcription Factor Inhibitors in Oncology
Edited by Khondaker Miraz Rahman and David E. Thurston

Published by the Royal Society of Chemistry, www.rsc.org

Increased stability of G-quartets is not limited to K^+ as other monovalent cations act similarly, however, early studies revealed that various monovalent cations had marked differences in guanosine gel stabilization.[4] Complex stability was linked to ionic radii of K^+, Rb^+, Na^+, Cs^+ and Li^+, and it was determined that the hydration energy of these cations was inversely proportional to their radii.[3,5] Out of this alkali series, K^+ produces the most stable G-quartet, particularly for intramolecular structures.

G-rich DNA can implement unusual structure formation, and this was first seen in denaturing polyacrylamide gel electrophoresis studies with telomeric sequences, and, later, NMR was used to verify the structure conformations. It was found with telomeric DNA that G-quartets fold in a way where the planar arrangements stack on top of each other with the interconnecting nucleotides forming loops connecting specific guanines. This marvel, now referred to as G-quadruplexes (G4s), was further complicated when more than one species was observed in a single DNA sequence. This polymorphic characteristic led to the deciphering of loop configurations as well as differences between intramolecular and intermolecular structures. Depending on the sequence, loop lengths can vary and control G4 topology depending on the direction that the loops adopt to connect guanine residues.[6] Loops link runs of continuous guanines at the top or bottom of G-quartet stacks; they can loop diagonally or laterally. If the loops connect guanines from the top to the bottom with a double-chain reversal, then the structure is in a parallel conformation. If the loops connect guanines diagonally or laterally, then the motif is in an antiparallel conformation. Hybrid structures are possible where some loops act in mixed directionalities; normally, G4s with short loops adopt parallel conformations. Another classification, intermolecular or intramolecular structures, depends on the DNA or RNA sequence in which is it being formed. If one strand participates in G4 formation, then it is an intramolecular assembly, while two or more strands organize into an intermolecular grouping.[5] From a biological standpoint, intramolecular structures are the ones of interest in promoter structures as potential targets for transcriptional regulation. Since several isoforms coexist *in vitro*, however, the predominant topology *in vivo* is challenging to predict.

7.1.1 Telomeres and G4s

The G-quartet structural motif is commonly found in telomeric DNA. Briefly, telomeres are located at the end of chromosomes and act as stabilizers of genomic integrity, are involved in DNA replication and help organize chromosomal DNA in the cell's nucleus. They can be single- or double-stranded DNA. Telomeres range in sequence length, but they tend to have several hundred thousand base pairs that are particularly rich in tandem guanine repeats. Telomeric DNA was of specific interest as many are extended single-strand sequences that allow for G-quadruplex formation. Evidence of these structures *in vivo*, obtained using specific antibodies, indicated the presence of an antiparallel intermolecular G4 at macronuclear telomeres within *Stylonychia*.[7] These secondary DNA structures were resolved during

DNA replication, which indicates that chromosomes may end in G4 caps. In humans, such structures form in telomeres and when unresolved, they can cause DNA damage by causing chromosomes to shorten. Specifically, G4-folding of the telomerase substrate has been examined and shown to impede catalytic enzyme activity.[8] Such inhibition leads to cell senescence and apoptosis as telomeric length rapidly shortens. With such extensive research surrounding telomeric DNA and the ability of G4s to form, it begged the question: could these structures form elsewhere in the genome?

7.1.2 Intracellular G4s Throughout the Genome

Extensive evidence describes the folding of non-B-DNA arrangements *in vitro* throughout the genome. Telomeric repeat-containing RNA (TERRAs) were characterized to form RNA-G4s within living human cells.[9] A pyrene excimer probe was utilized to observe unique fluorescent signaling to detect G4 arrangements in HeLa cells. Not only were RNA-G4s described in cells, but it was determined that such structures are localized at the ends of chromosomes from the HeLa cells, further supporting the prediction that RNA-G4s occur in telomeric regions of the human genome. A few years later, a DNA-G4 structure-specific antibody, BG4, was engineered to investigate the existence of G4 in human cells and within individual chromosomes.[10] G4s were visualized and shown to form both intracellularly within centromere and telomeric regions of DNA, as well as at origins of replication and in transcriptional and translational regulatory regions like promoters and untranslated regions (UTR's). Illumina next-generation sequencing of KCl- or compound-induced higher order structures identified around 700 000 G4s from genomic DNA, including both canonical and non-canonical putative G4 sequences. Notably, the majority of G4 locality was found to be broadly within regulatory regions and particularly within oncogenes.[11] These breakthroughs highlight the biological role of G4s in regulating gene expression and their therapeutic potential in cancer and other disease states.

Several computational methods have been utilized to quantify G4-forming potential throughout the genome. Overall, it is estimated that 40% of human gene promoters contain putative G4-forming sequences,[12] and studies have demonstrated that proto-oncogenes have very high G4-forming potential (Table 7.1), whereas tumor suppressor and housekeeping genes do not.[13] Guanine "richness" is associated with promoter regions with open chromatin structure where transcriptional activation is simpler. G4 formation occurs within these regions when DNA is unwound for transcription and when negative supercoiling occurs upstream from the transcriptional start site (TSS).[14]

Through their formation, G4s are predicted to disrupt regulatory protein binding. A number of G4s have been described for various oncogene promoters, including the cellular homolog of the myelocytomatosis viral oncogene (MYC), B-cell lymphoma 2 (Bcl-2), the cellular homolog of the rat sarcoma viral oncogene (RAS), the cellular homolog of the feline sarcoma viral oncogene (c-KIT), retinoblastoma (Rb), vascular endothelial growth factor (VEGF) and hypoxia-inducible factor 1-alpha (HIF-1α). While promoter

Table 7.1 G-quadruplexes regulating the hallmarks of cancer.

Hallmark	Example(s)	Data from Ref.
Angiogenesis	VEGF, VEGFR-2	148, 150
Invasion and Metastasis	c-KIT, ADAM-15	156–160
Tumor Promoting Inflammation	FGF-2	161
Replicative Immortality	Telomere, hTERT	3, 162–165
Avoiding Immune Destruction	TGF-β	166
Evasion of Growth Suppressors	pRb	167
Sustaining Proliferative Signaling	kRAS, MYC	16, 120, 123
Deregulation of Cellular Energetics	MYC, HIF-1α	16, 148
Resistance to Cell Death	Bcl-2	136–138
Genome Instability and Mutation	YY1	168

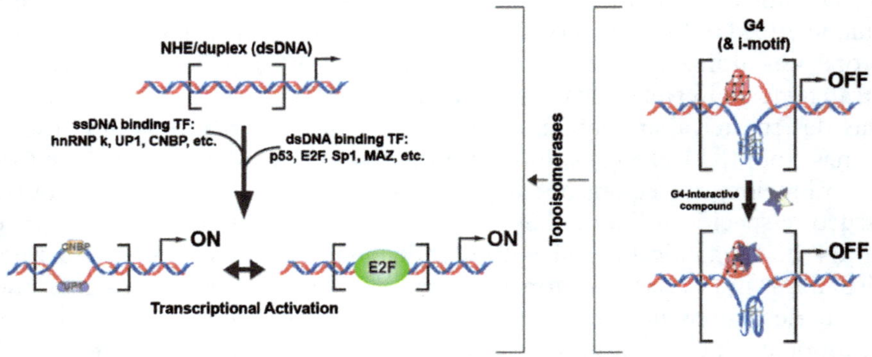

Figure 7.1 Dynamic regulation of nuclease G–C-rich nuclease hypersensitivity elements in transcriptionally active and silenced forms. Duplex DNA (top, left) can be activated by double-stranded (dsDNA)- or single-stranded (ssDNA)-binding transcription factors (TF), leading to gene expression. Negative superhelicitiy caused by transcription facilitates the formation of G4 and complementary strand i-motif structures (top, right), which sequester the binding sites for TF's and foster a transcriptionally inactive state that can be encouraged with the binding of G4 stabilizing compounds. Topoisomerases facilitate the resolution of higher-order G4 and i-motif structures by alleviating the torsional stress and shift the equilibrium of DNA back to a transcriptionally active state.

G4s tend to function as gene silencers, as described below with MYC, the function is individual to the promoter and dependent on the endogenous regulatory proteins (Figure 7.1). For example, G4 formation within the MYC promoter displaces the transcriptional activators specificity protein 1 (Sp1), cellular nucleic acid-binding protein (CNBP) and heterogeneous nuclear ribonucleoprotein K (hnRNP k), thus silencing gene expression,[15,16] whereas unique formation within the Bcl-2 promoter displaces the transcriptional repressor Wilms tumor protein 1 (WT1) and leads to enhanced expression.[17] Thus, G4 formation and stabilization is promising for alternative strategies in gene regulation for both upregulation and downregulation of oncogenes through transcriptional control.

7.1.3 G4 Complexity

Promoter G4s can be categorized into different structural classes. Simple secondary structures are considered to be Class I if one structural moiety predominates, perhaps having only slight differences within loop configurations with minor species; the MYC G4 is the best example of Class I formation. Class II G4s are represented in G-rich regions that can form two dissimilar structures that are separated by approximately three helical turns of DNA, as seen with c-Kit. The third class is similar in that it requires formation of two distinct structures, but the formations occur in tandem rather than separated by an extended region; the primary example of this class is found in the hTERT promoter where there are G4-related interstructural interactions. Lastly, class IV is differentiated from class I in that in a single G-rich region of DNA, more than one G4 can be formed from the same guanine-rich region. A 39 nucleotide long region within the Bcl-2 promoter consists of six runs of three or more consecutive guanine bases allowing for the formation of three G4s to exist.[14]

7.2 Drug Discovery Efforts *via* Ligand-based G4 Stabilization

7.2.1 A Historical Perspective

Maintenance of extended telomere regions is a critical hallmark of cancer and something necessary to maintain the increased proliferative rates that define the disease. Interference with telomere propagation has long been recognized as a viable anti-cancer therapeutic approach. With the discovery of G4-mediated telomeric damage through the inhibition of telomerase binding, the drug discovery efforts soon followed and still continue. The first G4-stabilizing compounds identified were large cyclic compounds such as the cationic porphyrin TMPyP4 [5,10,15,20-Tetrakis(1-methylpyridinium-4-yl)porphyrin tetra(*p*-toluenesulfonate)], a similar cationic diselenosapphyrin (Se2SAP) and the natural product macrocycle telomestatin.[18] Several other compound classes capable of stabilizing the telomeric G4 include the condensed hetercyclic core perylene compound PIPER (*N*-*N*′-bis[2-(1-piperidino)-ethyl]-3,4,9,10-perylenetetracarboxylic diimide)[19] and the ethidium/triazine derivatives PDC 360A (pyridodicarboxamide 360A),[20] Phen-DC_3 (phenanthroline derived compound 3)[21,22] and pyridostatin A.[8,23,24] When G4 studies extended into promoter regions, these were among the first compounds utilized to probe G4-mediated effects on transcription.

7.2.2 Mucking with MYC

MYC (also known as c-MYC, c-Myc and myc), a basic helix–loop–helix/leucine zipper transcription factor, is an enigmatic protein with over 30 000 potential binding sites in the human genome, 10–15% of which are generally bound at any one time. MYC has been shown to affect cellular proliferation, apoptosis, metastasis, angiogenesis and microenvironments.[25] Disregulated MYC has been noted in vasculogenesis,[26] restenosis,[27,28] genomic instability[29–32] and proteolysis,[33–36] but it is first and foremost known for its oncogenic role.[37–40] MYC was one of the first proto-oncogenes to be described and is ultimately disregulated in most tumor types and stages. In particular, MYC is overexpressed in up to 80% of all cancers, and a decrease in expression has significant therapeutic potential in many cancer types, including prostate, breast, leukemia and lymphoma.[25,38,40–50] As MYC is most often overexpressed through gene amplification or upregulation of transcription initiation, targeting gene expression through modulating promoter activity is a viable and promising therapeutic approach.

The role of dynamic DNA structural equilibrium in the transcriptional control of the proto-oncogene MYC was first described in 1987, in which a variation of S1 sensitivity was noted in proximity to the major transcription initiator P_1, herein termed the nuclease hypersensitivity region III_1, and 1500 bases upstream in an area now known as the FUSE (far upstream suppressor element) region.[51] Over a decade later, it was demonstrated that the alteration in footprinting sensitivity was due to higher-order DNA tetraplex structure formation in the nucleotide-hypersensitive element (NHE) III_1,[52] and soon thereafter that the transition from duplex to quadruplex DNA could be induced by the telomeric G4-stabilizing compound, PIPER[53] and that stabilization by the cationic porphyrin TMPyP4 suppressed MYC transcriptional activation.[54]

Several structures were proposed for the MYC promoter G4, including variations in loop directionality and composition[54–56] and inclusion of successive guanine runs,[54,57,58] but the predominant, physiologically relevant structure that is stabilized by TMPyP4 under torsional stress conditions has been demonstrated to be an all parallel structure incorporating the first four continuous runs of guanines, with intervening loops of one, two and one nucleotides from the 5′ to 3′ direction, respectively.[57] Extensive work has gone into describing the dynamic regulation of this region, which is known to be transcriptionally active in the duplex and single-stranded forms, as stimulated by Sp1 and CNBP, and inactive in the G-quadruplex structures form, as stabilized by nucleolin. nm23 H2 has been shown to facilitate the conversion from the higher-order non-canonical structure into the transcriptionally active single- and double-strand forms.[15,16]

As the first described promoter G-quadruplex structure, and given the high therapeutic value of MYC in cancer therapeutics, much work has gone into developing novel compounds capable of stabilizing the unique DNA structure using both small molecules and DNA-mediated techniques, described below. The classes of compounds capable of binding to, and stabilizing, the MYC promoter G-quadruplex include the aforementioned telomeric G-quadruplex-stabilizers,[59–61] other DNA dyes and natural products,[62–66] anticancer therapeutics, such as Actinomycin D, and an array of

FDA-approved drugs from many classes,[67–69] simple linear molecules, such as triethylene tetraamine,[70] metal complexes[71–78] and a variety of aromatic fused heterocycle compounds with five-[79–86] and six-member nitrogen-containing heterocycles,[87–95] or both together[96] (Figure 7.2).

From these compounds, anti-cancer activity *in vitro* and *in vivo* has been seen in leukemia,[83,97] lymphoma,[79] medulloblastoma,[98] cervical cancer,[99] nasopharyngeal cancer[100] and more. Particularly for the MYC promoter G-quadruplex, there exists a set of lymphoma cancer cell lines, RAMOS or RAJI and CA46, in which the MYC gene has been transposed to the immunoglobulin chain region, leading to marked transcriptional upregulation. These cell lines are notable for their retention (RAMOS and RAJI) or loss (CA46) of the NHE III_1 region and provide the basis for many molecular studies. The use of these cell lines has led to linking of the intracellular mechanism of action of a number of the aforementioned compounds to MYC promoter G-quadruplex stabilization.[16] However, the description of an allele-specific test in 2011[79] allowed for a more exacting measure of intracellular activity. This test, in particular, requires only the CA46 cell line and is able to differentiate between the intact G-quadruplex-maintaining minor allele with 1000-fold less transcriptional activity, *versus* the G-quadruplex-lacking major allele that is driven by the immunoglobulin enhancer region, as examined by both quantitative PCR (qPCR) and by chromatin immunoprecipitation (ChIP). This test was first used to describe the intracellular activity of NSC 338258 on the MYC promoter structure[79] and was later used to demonstrate that other compounds previously described, as well as a series of new quindoline compounds, did not suppress MYC in a G-quadruplex-specific manner.[101] While not all G-quadruplex-forming regions lend themselves to such detailed scrutiny, future compound development around this particular promoter structure is expected to be held to these standards.

It can be difficult to obtain significant selectivity for a particular promoter structure with small molecules; however, nucleic acids can be used to gain a high level of specificity. To this end, a number of nucleic-acid-based therapeutics have also been examined for the MYC promoter G-quadruplex. The approaches have been twofold as researchers have either chosen to directly apply the guanine-rich sequence of the MYC promoter to cells[102–104] or to complement the region with various probes[105–108] (Figure 7.3). The former method can lead to telomeric and non-telomeric cellular damage in leukemia and lymphoma cells,[102–104] ascribed to stabilization of promoter G-quadruplex structures including, but not limited to, MYC, as well as to disruption of telomeric protein structure stability. The latter approach demonstrates higher selectivity for MYC promoter activity, particularly when the DNA backbone is replaced with a locked or peptide nucleic acid structure,[105,107,108] which can lead to DNA scission.[106] While obstacles exist with the nucleic acid approach, such as cellular delivery and uptake, there are also marked advantages in specificity that enhance the attraction of developing novel formulations.

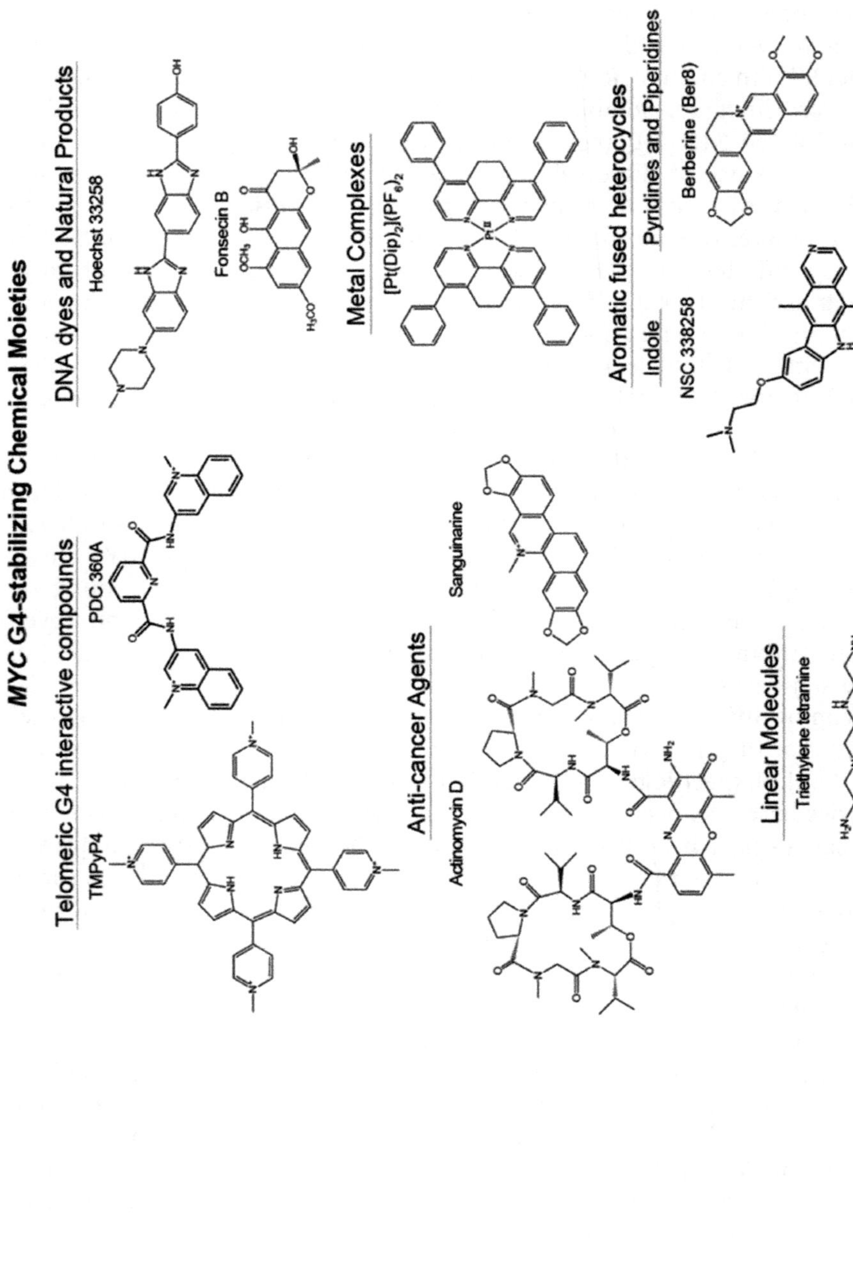

Figure 7.2 G4-interactive compounds shown to bind to the *MYC* promoter structure and decrease transcription.[56,60,63,64,68–70,74,79,90,91]

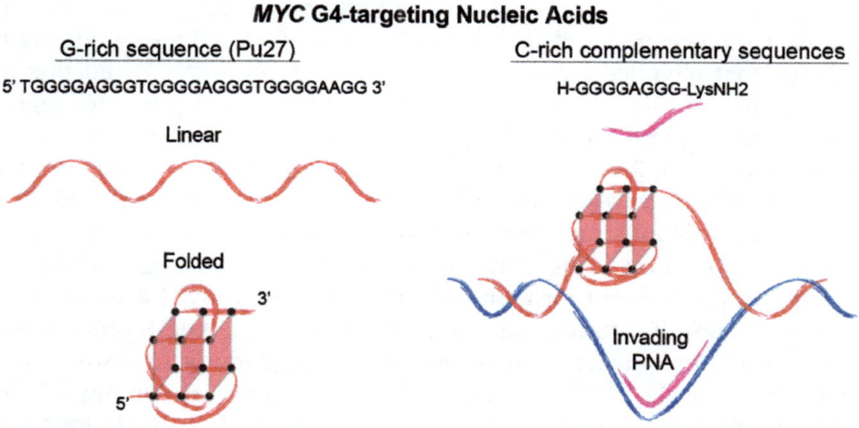

Figure 7.3 Nucleic-acid-mediated regulation of the *MYC* promoter G4. Application of the G4-forming sequence, Pu27, which is capable of being in either the linear or folded state intracellularly, facilitates leukemic cell death, both related and unrelated to MYC regulation.[103] Small modified sections of the G-rich region can also be applied as invading peptide nucleic acids (PNAs) that complement the C-rich strand of DNA, prevent the transition back to dsDNA and maintain a G4-state and subsequent transcriptional downregulation of MYC.[105]

7.2.3 The RAS Gene Family

The RAS gene family consists of three oncoproteins–Kirsten (k-), Harvey (H-) and Neuroblastoma (N-)RAS–each of which are responsible for encoding intracellular membrane-bound G-proteins involved in cell growth and apoptosis through the Raf–MAP kinase pathway.[109] The RAS genes are structurally similar and share high homology in exon sequences. Normally, the RAS protein toggles between a GDP-bound inactive and a GTP-bound active state. When active, RAS downstream signaling leads to cellular proliferation, metastasis and anti-apoptosis. Several point mutations can convert these genes into a constitutively active oncogenic form, causing aggressive cell growth. The most commonly mutated RAS protein is the k-isoform; this mutation is commonly seen in ovarian, lung, pancreatic and colon cancers.[110,111] Many attempts have been made to inactivate mutant RAS or affecting its downstream events, but none have developed into successful clinical agents, to date.[110] In contrast, downregulation of all k-RAS expression has demonstrated remarkable preclinical activity, but most approaches rely on nucleic acid approaches that require special formulation.[112–115] The ability to control k-RAS, or H- or N-RAS, expression with a small molecule would be of enormous clinical value.

The k-RAS core promoter is housed ±60 base pairs (bp) from the TSS; the proximal promoter region also contains a transcription enhancing region within the first 500 bp upstream of the TSS, within which lie two NHE regions consisting of three independent putative G4-forming sequences.[116–118]

Initial G4 characterization of the k-RAS promoter focused on the putative G-rich sequence closest to the TSS and described a unique parallel structure that included a discontinuous run of guanines.[119–122] Further delineation of the role of various G4s within the extended promoter region, however, highlighted the mid-G4-forming region as having the greatest potential to silence k-RAS expression.[123] This region forms a number of competing isoforms *ex vivo* and its *in vitro* structure is yet to be elucidated. Thus far, the predominant structure is noted to be a tri-stacked G-tetrad formation with mixed direction loops of two to ten nucleotides. The most distal G4-forming region does not form an inducible intramolecular G4 structure.[123]

Using TMPyP4, G4 formation was shown to decrease k-RAS promoter activity driven from luciferase plasmids and endogenously in PANC-1 pancreatic cancer cells.[109,124] Pyrene-modified G4 decoys were developed that stabilized the near-region k-RAS promoter G4 and significantly inhibited the viability of PANC-1 cells; its effect on the mid-region structure was not examined and remains unknown.[124] Locked nucleic acid G4-decoys of the near-G4 were also designed and were noted to decrease the metabolic activity of PANC-1 cells, to decrease k-RAS expression and to inhibit the growth of PANC-1 xenografts in severe combined immunodeficiency (SCID) mice;[125] the effect of these decoys on mid-G4 integrity was also not examined.

Small-molecule-mediated stabilization of k-RAS promoter has not been as extensively explored as that of MYC. In addition to the effects of TMPyP4, a series of indolo[3,2-b/c]quinolones demonstrated increased thermodynamic stability of the near-G4, downregulation of promoter activity and decreases in k-RAS mRNA expression in human colon cancer cells corresponding with apoptosis at 1–5 μM.[109,126] In the same fashion as the decoy-based therapeutic approaches, the effects of these compounds on the mid-G4 is not yet known. One other compound has been described as modulating the k-RAS promoter, a tetra-substituted naphthalene-diimide derivative (MM41) that stabilizes k-RAS1 and k-RAS2, which, although not detailed, are believed to represent the near- and mid-G4 formations, respectively. MM41 inhibited MIA PaCa-2 pancreatic cancer cell growth at nanomolar concentrations and decreased MIA PaCa-2 xenograft growth, with *in vivo* modulation of k-RAS expression. Notably, this compound also stabilized structures in the Bcl-2 promoter and modulated its expression, but did not effect telomere stability.[127] Overall, targeting G4 structures in the k-RAS promoter, particularly the mid-region formation, is a promising approach to therapeutic development for this previously "undruggable" target.

Mutant H-RAS frequently occurs in bladder cancers (about 80%).[128] Like k-RAS, this oncogene frequently harbors a point mutation causing constitutive activity.[129] Activation of H-RAS primarily correlates with tumor invasiveness, rather than the proliferative advantage described above. The HRAS promoter has two guanine-rich elements shown to form two G4s in tandem: hras-1 and hras-2. hras-1 prefers a parallel conformation while hras-2 prefers an antiparallel configuration; both are transcriptional silencers that can be stabilized with the addition of guanidinium phthalocyanines.[128]

Transcriptional silencing was also observed with the small molecule anthratiophenedione, which readily binds and stabilizes the HRAS G4s.[130] Intriguingly, if either G4 is mutated to disrupt the non-B-DNA formation, then the ability of the compound to modulate transcription was reduced by almost half, suggesting an equal contribution of each structure to transcriptional control.[131] More recently, chloroacetamidine–anthrathiophenedione derivatives have been designed and examined for G4 stabilization properties.[130] Lead analogs led to increased thermodynamic stability of the neighboring G4s, inhibited HRAS promoter activity and increased survival time of mice exposed to urinary bladder cancer xenografts.[132] Thus, a novel class of HRAS-selective G4-stabilizing compounds has been identified and shows promise in combating human bladder cancer.

N-RAS does harbor a targetable G4 in its regulatory region, within the 5′ UTR, that contributes to translational efficiency, rather than transcriptional control.[133,134] It is an attractive therapeutic target for melanoma and neuroblastoma, and it has been successfully targeted with pyridine-2,6-bis-quinolino-dicarboxamide derivatives, leading to decreased mRNA stability and reduced translation.[135]

7.2.4 Bcl-2: To Die or Not to Die?

Bcl-2 was named after its isolation from B-cell lymphoma, in which it suppressed apoptosis and caused oncogenesis. It is part of the BCL2 protein family that consists of apoptotic mediators. Bcl-2 encodes an anti-apoptotic mitochondrial membrane protein that aids in cell survival signaling by inhibiting the release of cytochrome c as well as activation of caspases. The human Bcl-2 gene has two promoters, P1 and P2, which are located 1300 base pairs apart. P1 is the predominant promoter and has numerous guanine-rich elements. Specifically, its 5′ end is involved in Bcl-2 transcription and regulation. The G-rich regions provide the foundation for the formation of G4s; three distinct guanine-rich regions have been demonstrated to form higher-order G4 structures located 19,[136] 176[137] and 451[138] bp upstream from the P1 start site and 1451, 1618 and 1875 bp upstream from the TSS. These structures are bcl2-G41, bcl2-G42 and bcl2-G43. Both bcl2-G42 and -G43 are putatively silencer elements, leading to a down-regulation of transcription, whereas the regulation by bcl2-G41 is inherently more complex. From this most proximal region, a number of competing G4 isoforms have been described, some with silencing potential,[139] some with transcriptional activating potential[140] and others with DNazyme/peroxidase potential.[141] The variation in function appears to be related to the endogenous transcription factors binding to the regions sequestered by G4-formation, particularly Sp1 and WT1. Sp1 is a predominantly a transcriptional activator, so its inability to bind the Bcl-2 promoter facilitates silencing and is ideal for oncological targeting. WT1, a product of the Wilms' tumor suppressor gene, binds approximately 27–35 base pairs upstream of the predominant Bcl-2 promoter and negatively regulates transcription. When non-canonical structures sequester this binding region, WT1 cannot bind and transcriptional activation is the result. Stabilization of the particular structure that sequesters WT1 binding sites, with small molecules, is advantageous when targeting neuronal diseases like Parkinson's disease, where apoptosis is occurring within dopaminergic neurons of the substantia nigra. However, for cancer the opposite effect is desirable, and specific stabilization of the Sp-1 sequestering bcl2-G41 structure, or of the -G42 or -G43 isoforms is the optimal therapeutic approach.

Bcl-2 is overexpressed in lymphomas, melanoma and myeloma. The Bcl-2 G4 has not been a collective choice for G4 drug discovery efforts, possibly due to the variable outcome of stabilizing each promoter G4, as described above. Antitumor agents, such as quindoline derivatives, are common G4-stabilizing agents and have been noted to decrease Bcl-2 promoter activity in plasmid studies.[139] However, quindolines tend to universally bind G4 structures,[101,142–144] and more selective derivatives will need to be engineered for further drug discovery, particularly for the Bcl-2 promoter. Bcl-2 G4 selectivity has been achieved with dinuclear ruthenium(II) complexes that bind G4s and force configurational changes in DNA. In HeLa cells, apoptosis was induced, but intracellular gene regulation was undescribed.[145] bcl2-G42 stability has been targeted with peptide nucleic acids (PNA's),[137,146] but not yet with small molecules; their biological activity also remains to be described. Regulating the Bcl-2 promoter, or the structure in the 5′ UTR that dampens translation,[147] *via* higher order DNA structures, is a good candidate approach for future chemotherapeutics for liquid tumors if structure selectivity were achievable.

7.3 Where Do We Go from Here?

There are a number of other promoters with well-described promoter secondary DNA structures as well, such as c-KIT, which contains two distinct promoter G-quadruplex-forming regions, VEGF, VEGF receptor 2 (VEGFR2), platelet-derived growth factor receptor alpha (PDGFR-α), platelet-derived growth factor receptor beta (PDGFR-β), HIF-1α and many more.[14,148–151] The targeted stabilization of these structures is developing almost as rapidly as their descriptions, and the literature is expanding at an exponential rate. The ability to modulate transcription by targeting DNA in a manner that is more specific than major- or minor-groove interactions, or by intercalation, is an attractive therapeutic possibility. With the recent expansion of descriptions of G-quadruplexes in whole cells through antibodies, small molecules and fluorescent probes, the field is expected to continue to grow. There are two cases of G-quadruplex-targeted therapeutics reaching the stage of clinical trials: Quarfloxin, which was stopped at phase II due to excessive albumin binding, and AS1411, which was well tolerated but offered less than optimal efficacy. The former is currently being examined by formulation and analoging, and the latter is under investigation in nano-delivery vehicles, including as gold conjugates.[152] It is clear that as G-quadruplex-targeted compounds for transcriptional control are developed, more stringent experimental conditions are required to convincingly demonstrate intracellular mechanisms of action. Such assays could include the exon-specific assay described above for MYC,[79] the use of fluorescent dyes, or the combination of a G-quadruplex antibody, such as BG4, post compound application.[11,153–155] G4 research continues to grow as evidence indicates that they exist throughout the genome. Equipped with this compelling data, the formation of G4 drug discovery programs to develop efficacious small molecules shows therapeutic promise for many critical disease states, most notably in oncology.

References

1. I. Bang, Untersuchungen über die Guanylsäure, *Biochem. Z.*, 1910, **26**, 293–311.
2. M. Gellert, M. N. Lipsett and D. R. Davies, Helix formation by guanylic acid, *Proc. Natl. Acad. Sci. U. S. A.*, 1962, **48**, 2013–2018.
3. J. R. Williamson, G-quartet structures in telomeric DNA, *Annu. Rev. Biophys. Biomol. Struct.*, 1994, **23**, 703–730.
4. W. Guschlbauer, J. F. Chantot and D. Thiele, Four-stranded nucleic acid structures 25 years later: from guanosine gels to telomer DNA, *J. Biomol. Struct. Dyn.*, 1990, **8**, 491–511.
5. T. M. Bryan and P. Baumann, G-quadruplexes: from guanine gels to chemotherapeutics, *Mol. Biotechnol.*, 2011, **49**, 198–208.
6. J. L. Huppert, Structure, location and interactions of G-quadruplexes, *FEBS J.*, 2010, **277**, 3452–3458.
7. C. Schaffitzel, I. Berger, J. Postberg, J. Hanes, H. J. Lipps and A. Pluckthun, In vitro generated antibodies specific for telomeric guanine-quadruplex DNA react with Stylonychia lemnae macronuclei, *Proc. Natl. Acad. Sci. U. S. A.*, 2001, **98**, 8572–8577.
8. S. Neidle, Human telomeric G-quadruplex: the current status of telomeric G-quadruplexes as therapeutic targets in human cancer, *FEBS J.*, 2010, **277**, 1118–1125.
9. Y. Xu, Y. Suzuki, K. Ito and M. Komiyama, Telomeric repeat-containing RNA structure in living cells, *Proc. Natl. Acad. Sci. U. S. A.*, 2010, **107**, 14579–14584.
10. G. Biffi, D. Tannahill, J. McCafferty and S. Balasubramanian, Quantitative visualization of DNA G-quadruplex structures in human cells, *Nat. Chem.*, 2013, **5**, 182–186.
11. V. S. Chambers, G. Marsico, J. M. Boutell, M. Di Antonio, G. P. Smith and S. Balasubramanian, High-throughput sequencing of DNA G-quadruplex structures in the human genome, *Nat. Biotechnol.*, 2015, **33**, 877–881.
12. J. L. Huppert and S. Balasubramanian, G-quadruplexes in promoters throughout the human genome, *Nucleic Acids Res.*, 2007, **35**, 406–413.

13. J. Eddy and N. Maizels, Gene function correlates with potential for G4 DNA formation in the human genome, *Nucleic Acids Res.*, 2006, **34**, 3887–3896.
14. T. A. Brooks, S. Kendrick and L. Hurley, Making sense of G-quadruplex and i-motif functions in oncogene promoters, *FEBS J.*, 2010, **277**, 3459–3469.
15. T. A. Brooks and L. H. Hurley, The role of supercoiling in transcriptional control of MYC and its importance in molecular therapeutics, *Nat. Rev.*, 2009, **9**, 849–861.
16. T. A. Brooks and L. H. Hurley, Targeting MYC Expression through G-Quadruplexes, *Genes Cancer*, 2010, **1**, 641–649.
17. S. Kendrick, H. J. Kang, M. P. Alam, M. M. Madathil, P. Agarwal, V. Gokhale, D. Yang, S. M. Hecht and L. H. Hurley, The dynamic character of the BCL2 promoter i-motif provides a mechanism for modulation of gene expression by compounds that bind selectively to the alternative DNA hairpin structure, *J. Am. Chem. Soc.*, 2014, **136**(11), 4161–4171.
18. E. M. Rezler, J. Seenisamy, S. Bashyam, M. Y. Kim, E. White, W. D. Wilson and L. H. Hurley, Telomestatin and diseleno sapphyrin bind selectively to two different forms of the human telomeric G-quadruplex structure, *J. Am. Chem. Soc.*, 2005, **127**, 9439–9447.
19. R. T. Pillich, G. Scarsella, G. Galati, L. Izzo, A. Iacoangeli, M. Castelli and G. Risuleo, The diimide drug PIPER has a cytotoxic dose-dependent effect in vitro and inhibits telomere elongation in HELA cells, *Anticancer Res.*, 2005, **25**, 3341–3346.
20. C. Granotier, G. Pennarun, L. Riou, F. Hoffschir, L. R. Gauthier, A. De Cian, D. Gomez, E. Mandine, J. F. Riou, J. L. Mergny, P. Mailliet, B. Dutrillaux and F. D. Boussin, Preferential binding of a G-quadruplex ligand to human chromosome ends, *Nucleic Acids Res.*, 2005, **33**, 4182–4190.
21. A. De Cian, E. Delemos, J. L. Mergny, M. P. Teulade-Fichou and D. Monchaud, Highly efficient G-quadruplex recognition by bisquinolinium compounds, *J. Am. Chem. Soc.*, 2007, **129**, 1856–1857.
22. A. Piazza, J. B. Boule, J. Lopes, K. Mingo, E. Largy, M. P. Teulade-Fichou and A. Nicolas, Genetic instability triggered by G-quadruplex interacting Phen-DC compounds in Saccharomyces cerevisiae, *Nucleic Acids Res.*, 2010, **38**, 4337–4348.
23. T. Vy Thi Le, S. Han, J. Chae and H. J. Park, G-quadruplex binding ligands: from naturally occurring to rationally designed molecules, *Curr. Pharm. Des.*, 2012, **18**, 1948–1972.
24. S. Zhang, Y. Wu and W. Zhang, G-quadruplex structures and their interaction diversity with ligands, *ChemMedChem*, 2014, **9**, 899–911.
25. C. V. Dang, Enigmatic MYC conducts an Unfolding Systems Biology Symphoma, *Genes Cancer*, 2010, **1**, 526–531.
26. T. A. Baudino, C. McKay, H. Pendeville-Samain, J. A. Nilsson, K. H. Maclean, E. L. White, A. C. Davis, J. N. Ihle and J. L. Cleveland,

c-Myc is essential for vasculogenesis and angiogenesis during development and tumor progression, *Genes Dev.*, 2002, **16**, 2530–2543.
27. C. Bauters, P. de Groote, M. Adamantidis, C. Delcayre, M. Hamon, J. M. Lablanche, M. E. Bertrand, B. Dupuis and B. Swynghedauw, Proto-oncogene expression in rabbit aorta after wall injury, First marker of the cellular process leading to restenosis after angioplasty?, *Eur. Heart J.*, 1992, **13**, 556–559.
28. S. Biro, Y. M. Fu, Z. X. Yu and S. E. Epstein, Inhibitory effects of antisense oligodeoxynucleotides targeting c-myc mRNA on smooth muscle cell proliferation and migration, *Proc. Natl. Acad. Sci. U. S. A.*, 1993, **90**, 654–658.
29. S. Mai, M. Fluri, D. Siwarski and K. Huppi, Genomic instability in MycER-activated Rat1A-MycER cells, *Chromosome Res.*, 1996, **4**, 365–371.
30. C. Taylor, A. Jalava and S. Mai, c-Myc dependent initiation of genomic instability during neoplastic transformation, *Curr. Top. Microbiol. Immunol.*, 1997, **224**, 201–207.
31. C. Taylor and S. Mai, c-Myc-associated genomic instability of the dihydrofolate reductase locus in vivo, *Cancer Detect. Prev.*, 1998, **22**, 350–356.
32. X. Y. Yin, L. Grove, N. S. Datta, M. W. Long and E. V. Prochownik, C-myc overexpression and p53 loss cooperate to promote genomic instability, *Oncogene*, 1999, **18**, 1177–1184.
33. D. H. Crouch, V. J. Fincham and M. C. Frame, Targeted proteolysis of the focal adhesion kinase pp125 FAK during c-MYC-induced apoptosis is suppressed by integrin signalling, *Oncogene*, 1996, **12**, 2689–2696.
34. R. Gavioli, T. Frisan, S. Vertuani, G. W. Bornkamm and M. G. Masucci, c-myc overexpression activates alternative pathways for intracellular proteolysis in lymphoma cells, *Nat. Cell Biol.*, 2001, **3**, 283–288.
35. S. J. Martin, G. A. O'Brien, W. K. Nishioka, A. J. McGahon, A. Mahboubi, T. C. Saido and D. R. Green, Proteolysis of fodrin (non-erythroid spectrin) during apoptosis, *J. Biol. Chem.*, 1995, **270**, 6425–6428.
36. R. C. O'Hagan, M. Ohh, G. David, I. M. de Alboran, F. W. Alt, W. G. Kaelin, Jr. and R. A. DePinho, Myc-enhanced expression of Cul1 promotes ubiquitin-dependent proteolysis and cell cycle progression, *Genes Dev.*, 2000, **14**, 2185.
37. F. E. Cotter and E. Zucca, Altered gene expression and oncogenesis of B-cell neoplasia, *Ann. Oncol.*, 1991, **2**, 335–342.
38. G. J. Kato and C. V. Dang, Function of the c-Myc oncoprotein, *FASEB J.*, 1992, **6**, 3065–3072.
39. W. Lutz, J. Leon and M. Eilers, Contributions of Myc to tumorigenesis, *Biochim. Biophys. Acta*, 2002, **1602**, 61–71.
40. C. E. Nesbit, J. M. Tersak and E. V. Prochownik, MYC oncogenes and human neoplastic disease, *Oncogene*, 1999, **18**, 3004–3016.
41. H. J. Chung and D. Levens, c-myc expression: keep the noise down!, *Mol. Cells*, 2005, **20**, 157–166.

42. S. Cory, S. Gerondakis, L. M. Corcoran, O. Bernard, E. Webb and J. M. Adams, Activation of the c-myc oncogene in B and T lymphoid tumors, *Curr. Top. Microbiol. Immunol.*, 1984, **113**, 161–165.
43. C. V. Dang and L. A. Lee, *c-Myc Function in Neoplasia*, R.G. Landes Co, Place Published, 1995.
44. C. V. Dang, K. A. O'Donnell, K. I. Zeller, T. Nguyen, R. C. Osthus and F. Li, The c-Myc target gene network, *Semin. Cancer Biol.*, 2006, **16**, 253–264.
45. M. Eilers and R. N. Eisenman, Myc's broad reach, *Genes Dev.*, 2008, **22**, 2755–2766.
46. S. Janz, Myc translocations in B cell and plasma cell neoplasms, *DNA Repair*, 2006, **5**, 1213–1224.
47. K. Kelly and U. Siebenlist, The role of c-myc in the proliferation of normal and neoplastic cells, *J. Clin. Immunol.*, 1985, **5**, 65–77.
48. C. M. Koh, C. J. Bieberich, C. V. Dang, W. G. Nelson, S. Yegnasubramanian and A. M. De Marzo, MYC and Prostate Cancer, *Genes Cancer*, 2010, **1**, 617–628.
49. D. J. Liao and R. B. Dickson, c-Myc in breast cancer, *Endocr.-Relat. Cancer*, 2000, **7**, 143–164.
50. N. Meyer and L. Z. Penn, Reflecting on 25 years with MYC, *Nat. Rev. Cancer*, 2008, **8**, 976–990.
51. T. C. Boles and M. E. Hogan, DNA structure equilibria in the human c-myc gene, *Biochemistry*, 1987, **26**, 367–376.
52. T. Simonsson, P. Pecinka and M. Kubista, DNA tetraplex formation in the control region of c-myc, *Nucleic Acids Res.*, 1998, **26**, 1167–1172.
53. A. Rangan, O. Y. Fedoroff and L. H. Hurley, Induction of duplex to G-quadruplex transition in the c-myc promoter region by a small molecule, *J. Biol. Chem.*, 2001, **276**, 4640–4646.
54. A. Siddiqui-Jain, C. L. Grand, D. J. Bearss and L. H. Hurley, Direct evidence for a G-quadruplex in a promoter region and its targeting with a small molecule to repress c-MYC transcription, *Proc. Natl. Acad. Sci. U. S. A.*, 2002, **99**, 11593–11598.
55. A. T. Phan, Y. S. Modi and D. J. Patel, Propeller-type parallel-stranded G-quadruplexes in the human c-myc promoter, *J. Am. Chem. Soc.*, 2004, **126**, 8710–8716.
56. J. Seenisamy, E. M. Rezler, T. J. Powell, D. Tye, V. Gokhale, C. S. Joshi, A. Siddiqui-Jain and L. H. Hurley, The dynamic character of the G-quadruplex element in the c-MYC promoter and modification by TMPyP4, *J. Am. Chem. Soc.*, 2004, **126**, 8702–8709.
57. D. Sun and L. H. Hurley, The importance of negative superhelicity in inducing the formation of G-quadruplex and i-motif structures in the c-Myc promoter: implications for drug targeting and control of gene expression, *J. Med. Chem.*, 2009, **52**, 2863–2874.
58. D. Yang and L. H. Hurley, Structure of the biologically relevant G-quadruplex in the c-MYC promoter, *Nucleosides, Nucleotides nucleic Acids*, 2006, **25**, 951–968.

59. V. Gabelica, E. S. Baker, M. P. Teulade-Fichou, E. De Pauw and M. T. Bowers, Stabilization and structure of telomeric and c-myc region intramolecular G-quadruplexes: the role of central cations and small planar ligands, *J. Am. Chem. Soc.*, 2007, **129**, 895–904.
60. T. Lemarteleur, D. Gomez, R. Paterski, E. Mandine, P. Mailliet and J. F. Riou, Stabilization of the c-myc gene promoter quadruplex by specific ligands' inhibitors of telomerase, *Biochem. Biophys. Res. Commun.*, 2004, **323**, 802–808.
61. J. Seenisamy, S. Bashyam, V. Gokhale, H. Vankayalapati, D. Sun, A. Siddiqui-Jain, N. Streiner, K. Shin-ya, E. White, W. D. Wilson and L. H. Hurley, Design and synthesis of an expanded porphyrin that has selectivity for the c-MYC G-quadruplex structure, *J. Am. Chem. Soc.*, 2005, **127**, 2944–2959.
62. D. S. Chan, H. Yang, M. H. Kwan, Z. Cheng, P. Lee, L. P. Bai, Z. H. Jiang, C. Y. Wong, W. F. Fong, C. H. Leung and D. L. Ma, Structure-based optimization of FDA-approved drug methylene blue as a c-myc G-quadruplex DNA stabilizer, *Biochimie*, 2011, **93**, 1055–1064.
63. H. M. Lee, D. S. Chan, F. Yang, H. Y. Lam, S. C. Yan, C. M. Che, D. L. Ma and C. H. Leung, Identification of natural product fonsecin B as a stabilizing ligand of c-myc G-quadruplex DNA by high-throughput virtual screening, *Chem. Commun.*, 2010, **46**, 4680–4682.
64. S. Maiti, N. K. Chaudhury and S. Chowdhury, Hoechst 33258 binds to G-quadruplex in the promoter region of human c-myc, *Biochem. Biophys. Res. Commun.*, 2003, **310**, 505–512.
65. N. Ranjan and D. P. Arya, Targeting C-myc G-quadruplex: dual recognition by aminosugar-bisbenzimidazoles with varying linker lengths, *Molecules*, 2013, **18**, 14228–14240.
66. M. Tian, X. Zhang, Y. Li, Y. Ju, J. Xiang, C. Zhao and Y. Tang, Inducement of G-quadruplex DNA forming and down-regulation of oncogene c-myc by bile acid-amino acid conjugate-BAA, *Nucleosides, Nucleotides Nucleic Acids*, 2010, **29**, 190–199.
67. D. Castillo-Gonzalez, G. Perez-Machado, A. Guedin, J. L. Mergny and M. A. Cabrera-Perez, FDA-approved drugs selected using virtual screening bind specifically to G-quadruplex DNA, *Curr. Pharm. Des.*, 2013, **19**, 2164–2173.
68. S. Ghosh, S. K. Pradhan, A. Kar, S. Chowdhury and D. Dasgupta, Molecular basis of recognition of quadruplexes human telomere and c-myc promoter by the putative anticancer agent sanguinarine, *Biochim. Biophys. Acta*, 2013, **1830**, 4189–4201.
69. H. J. Kang and H. J. Park, Novel molecular mechanism for actinomycin D activity as an oncogenic promoter G-quadruplex binder, *Biochemistry*, 2009, **48**, 7392–7398.
70. F. Yin, J. Liu, X. Deng and J. Wang, Effects of triethylene tetraamine on the G-quadruplex structure in the human c-myc promoter, *J. Biochem.*, 2007, **141**, 669–674.

71. J. Alzeer and N. W. Luedtke, pH-mediated fluorescence and G-quadruplex binding of amido phthalocyanines, *Biochemistry*, 2010, **49**, 4339–4348.
72. N. P. Barry, N. H. Abd Karim, R. Vilar and B. Therrien, Interactions of ruthenium coordination cubes with DNA, *Dalton Trans.*, 2009, 10717–10719.
73. K. Suntharalingam, A. J. White and R. Vilar, Two metals are better than one: investigations on the interactions between dinuclear metal complexes and quadruplex DNA, *Inorg. Chem.*, 2010, **49**, 8371–8380.
74. J. Wang, K. Lu, S. Xuan, Z. Toh, D. Zhang and F. Shao, A Pt(II)-Dip complex stabilizes parallel c-myc G-quadruplex, *Chem. Commun.*, 2013, **49**, 4758–4760.
75. P. Wang, C. H. Leung, D. L. Ma, S. C. Yan and C. M. Che, Structure-based design of platinum(II) complexes as c-myc oncogene down-regulators and luminescent probes for G-quadruplex DNA, *Chemistry*, 2010, **16**, 6900–6911.
76. C. Wei, Y. Wen and J. Wang, Novel platinum complexes as efficient G-quadruplex DNA binders and telomerase inhibitors, *Int. J. Biol. Macromol.*, 2013, **55**, 185–192.
77. H. Yang, V. P. Ma, D. S. Chan, H. Z. He, C. H. Leung and D. L. Ma, A cyclometallated iridium(III) complex as a c-myc G-quadruplex stabilizer and down-regulator of c-myc oncogene expression, *Curr. Med. Chem.*, 2013, **20**, 576–582.
78. Z. Zhang, Q. Wu, X. H. Wu, F. Y. Sun, L. M. Chen, J. C. Chen, S. L. Yang and W. J. Mei, Ruthenium(II) complexes as apoptosis inducers by stabilizing c-myc G-quadruplex DNA, *Eur. J. Med. Chem.*, 2014, **80**, 316–324.
79. R. V. Brown, F. L. Danford, V. Gokhale, L. H. Hurley and T. A. Brooks, Demonstration that drug-targeted down-regulation of MYC in non-Hodgkins lymphoma is directly mediated through the promoter G-quadruplex, *J. Biol. Chem.*, 2011, **286**, 41018–41027.
80. W. J. Chen, C. X. Zhou, P. F. Yao, X. X. Wang, J. H. Tan, D. Li, T. M. Ou, L. Q. Gu and Z. S. Huang, Disubstituted 1,8-dipyrazolcarbazole derivatives as a new type of c-myc G-quadruplex binding ligands, *Bioorg. Med. Chem.*, 2012, **20**, 2829–2836.
81. J. Dash, P. S. Shirude and S. Balasubramanian, G-quadruplex recognition by bis-indole carboxamides, *Chem. Commun.*, 2008, 3055–3057.
82. D. N. Kaluzhny, A. K. Shchyolkina, N. S. Ilyinsky, O. F. Borisova and A. A. Shtil, Novel Indolocarbazole Derivative 12-(alpha-L-arabinopyranosyl)indolo[2,3-a]pyrrolo[3,4-c]carbazole-5,7-dione Is a Preferred c-Myc Guanine Quadruplex Ligand, *J. Nucleic Acids*, 2011, **2011**, 184735.
83. J. N. Liu, R. Deng, J. F. Guo, J. M. Zhou, G. K. Feng, Z. S. Huang, L. Q. Gu, Y. X. Zeng and X. F. Zhu, Inhibition of myc promoter and telomerase activity and induction of delayed apoptosis by SYUIQ-5, a novel G-quadruplex interactive agent in leukemia cells, *Leukemia*, 2007, **21**, 1300–1302.

84. D. L. Ma, D. S. Chan, W. C. Fu, H. Z. He, H. Yang, S. C. Yan and C. H. Leung, Discovery of a natural product-like c-myc G-quadruplex DNA groove-binder by molecular docking, *PLoS One*, 2012, **7**, e43278.
85. H. S. Mohammed, J. O. Delos Santos and B. A. Armitage, Noncovalent binding and fluorogenic response of cyanine dyes to DNA homoquadruplex and PNA-DNA heteroquadruplex structures, *Artificial DNA, PNA XNA*, 2011, **2**, 43–49.
86. T. M. Ou, Y. J. Lu, C. Zhang, Z. S. Huang, X. D. Wang, J. H. Tan, Y. Chen, D. L. Ma, K. Y. Wong, J. C. Tang, A. S. Chan and L. Q. Gu, Stabilization of G-quadruplex DNA and down-regulation of oncogene c-myc by quindoline derivatives, *J. Med. Chem.*, 2007, **50**, 1465–1474.
87. H. Bertrand, A. Granzhan, D. Monchaud, N. Saettel, R. Guillot, S. Clifford, A. Guedin, J. L. Mergny and M. P. Teulade-Fichou, Recognition of G-quadruplex DNA by triangular star-shaped compounds: with or without side chains?, *Chemistry*, 2011, **17**, 4529–4539.
88. X. Ji, H. Sun, H. Zhou, J. Xiang, Y. Tang and C. Zhao, The interaction of telomeric DNA and C-myc22 G-quadruplex with 11 natural alkaloids, *Nucleic Acid Ther.*, 2012, **22**, 127–136.
89. S. R. Liao, C. X. Zhou, W. B. Wu, T. M. Ou, J. H. Tan, D. Li, L. Q. Gu and Z. S. Huang, 12-N-Methylated 5,6-dihydrobenzo[c]acridine derivatives: a new class of highly selective ligands for c-myc G-quadruplex DNA, *Eur. J. Med. Chem.*, 2012, **53**, 52–63.
90. Y. Ma, T. M. Ou, J. Q. Hou, Y. J. Lu, J. H. Tan, L. Q. Gu and Z. S. Huang, 9-N-Substituted berberine derivatives: stabilization of G-quadruplex DNA and down-regulation of oncogene c-myc, *Bioorg. Med. Chem.*, 2008, **16**, 7582–7591.
91. Y. Ma, T. M. Ou, J. H. Tan, J. Q. Hou, S. L. Huang, L. Q. Gu and Z. S. Huang, Quinolino-benzo-[5, 6]-dihydroisoquindolium compounds derived from berberine: a new class of highly selective ligands for G-quadruplex DNA in c-myc oncogene, *Eur. J. Med. Chem.*, 2011, **46**, 1906–1913.
92. C. Pivetta, L. Lucatello, A. P. Krapcho, B. Gatto, M. Palumbo and C. Sissi, Perylene side chains modulate G-quadruplex conformation in biologically relevant DNA sequences, *Bioorg. Med. Chem.*, 2008, **16**, 9331–9339.
93. C. Wei, Y. Wang and M. Zhang, Synthesis and binding studies of novel di-substituted phenanthroline compounds with genomic promoter and human telomeric DNA G-quadruplexes, *Org. Biomol. Chem.*, 2013, **11**, 2355–2364.
94. H. Yang, H. J. Zhong, K. H. Leung, D. S. Chan, V. P. Ma, W. C. Fu, R. Nanjunda, W. D. Wilson, D. L. Ma and C. H. Leung, Structure-based design of flavone derivatives as c-myc oncogene down-regulators, *Eur. J. Pharm. Sci.*, 2013, **48**, 130–141.
95. Z. Zhang, X. He and G. Yuan, Regulation of the equilibrium between G-quadruplex and duplex DNA in promoter of human c-myc oncogene by a pyrene derivative, *Int. J. Biol. Macromol.*, 2011, **49**, 1173–1176.

96. F. C. Huang, C. C. Chang, J. M. Wang, T. C. Chang and J. J. Lin, Induction of senescence in cancer cells by the G-quadruplex stabilizer, BMVC4, is independent of its telomerase inhibitory activity, *Br. J. Pharmacol.*, 2012, **167**, 393–406.
97. Y. Mikami-Terao, M. Akiyama, Y. Yuza, T. Yanagisawa, O. Yamada and H. Yamada, Antitumor activity of G-quadruplex-interactive agent TMPyP4 in K562 leukemic cells, *Cancer Lett.*, 2008, **261**, 226–234.
98. T. Shalaby, A. O. von Bueren, M. L. Hurlimann, G. Fiaschetti, D. Castelletti, T. Masayuki, K. Nagasawa, A. Arcaro, I. Jelesarov, K. Shin-ya and M. Grotzer, Disabling c-Myc in childhood medulloblastoma and atypical teratoid/rhabdoid tumor cells by the potent G-quadruplex interactive agent S2T1-6OTD, *Mol. Cancer Ther.*, 2010, **9**, 167–179.
99. W. J. Zhou, R. Deng, X. Y. Zhang, G. K. Feng, L. Q. Gu and X. F. Zhu, G-quadruplex ligand SYUIQ-5 induces autophagy by telomere damage and TRF2 delocalization in cancer cells, *Mol. Cancer Ther.*, 2009, **8**, 3203–3213.
100. W. J. Zhou, R. Deng, G. K. Feng and X. F. Zhu, [A G-quadruplex ligand SYUIQ-5 induces autophagy by inhibiting the Akt-FOXO3a pathway in nasopharyngeal cancer cells], *Aizheng*, 2009, **28**, 1049–1053.
101. P. V. Boddupally, S. Hahn, C. Beman, B. De, T. A. Brooks, V. Gokhale and L. H. Hurley, Anticancer activity and cellular repression of c-MYC by the G-quadruplex-stabilizing 11-piperazinylquindoline is not dependent on direct targeting of the G-quadruplex in the c-MYC promoter, *J. Med. Chem.*, 2012, **55**, 6076–6086.
102. M. A. Islam, S. D. Thomas, V. V. Murty, K. J. Sedoris and D. M. Miller, c-Myc quadruplex-forming sequence Pu-27 induces extensive damage in both telomeric and nontelomeric regions of DNA, *J. Biol. Chem.*, 2014, **289**, 8521–8531.
103. K. C. Sedoris, S. D. Thomas, C. R. Clarkson, D. Muench, A. Islam, R. Singh and D. M. Miller, Genomic c-Myc quadruplex DNA selectively kills leukemia, *Mol. Cancer Ther.*, 2012, **11**, 66–76.
104. T. Simonsson and M. Henriksson, c-myc Suppression in Burkitt's lymphoma cells, *Biochem. Biophys. Res. Commun.*, 2002, **290**, 11–15.
105. A. Gupta, L. L. Lee, S. Roy, F. A. Tanious, W. D. Wilson, D. H. Ly and B. A. Armitage, Strand invasion of DNA quadruplexes by PNA: comparison of homologous and complementary hybridization, *ChemBioChem*, 2013, **14**, 1476–1484.
106. T. Ishizuka, J. Yang, M. Komiyama and Y. Xu, G-rich sequence-specific recognition and scission of human genome by PNA/DNA hybrid G-quadruplex formation, *Angew. Chem.*, 2012, **51**, 7198–7202.
107. N. Kumar, A. Patowary, S. Sivasubbu, M. Petersen and S. Maiti, Silencing c-MYC expression by targeting quadruplex in P1 promoter using locked nucleic acid trap, *Biochemistry*, 2008, **47**, 13179–13188.
108. S. Roy, F. A. Tanious, W. D. Wilson, D. H. Ly and B. A. Armitage, High-affinity homologous peptide nucleic acid probes for targeting a

quadruplex-forming sequence from a MYC promoter element, *Biochemistry*, 2007, **46**, 10433–10443.
109. J. Lavrado, H. Brito, P. M. Borralho, S. A. Ohnmacht, N. S. Kim, C. Leitao, S. Pisco, M. Gunaratnam, C. M. Rodrigues, R. Moreira, S. Neidle and A. Paulo, KRAS oncogene repression in colon cancer cell lines by G-quadruplex binding indolo[3,2-c]quinolines, *Sci. Rep.*, 2015, **5**, 9696.
110. A. A. Adjei, Blocking oncogenic Ras signaling for cancer therapy, *J. Natl. Cancer Inst.*, 2001, **93**, 1062–1074.
111. A. Young, J. Lyons, A. L. Miller, V. T. Phan, I. R. Alarcon and F. McCormick, Ras signaling and therapies, *Adv. Cancer Res.*, 2009, **102**, 1–17.
112. A. M. Duursma and R. Agami, Ras interference as cancer therapy, *Semin. Cancer Biol.*, 2003, **13**, 267–273.
113. B. B. Friday and A. A. Adjei, K-ras as a target for cancer therapy, *Biochim. Biophys. Acta*, 2005, **1756**, 127–144.
114. K. Podsypanina, K. Politi, L. J. Beverly and H. E. Varmus, Oncogene cooperation in tumor maintenance and tumor recurrence in mouse mammary tumors induced by Myc and mutant Kras, *Proc. Natl. Acad. Sci. U. S. A.*, 2008, **105**, 5242–5247.
115. E. Wickstrom, Oligonucleotide treatment of ras-induced tumors in nude mice, *Mol. Biotechnol.*, 2001, **18**, 35–55.
116. J. Jordano and M. Perucho, Chromatin structure of the promoter region of the human c-K-ras gene, *Nucleic Acids Res.*, 1986, **14**, 7361–7378.
117. J. Jordano and M. Perucho, Initial characterization of a potential transcriptional enhancer for the human c-K-ras gene, *Oncogene*, 1988, **2**, 359–366.
118. F. Yamamoto and M. Perucho, Characterization of the human c-K-ras gene promoter, *Oncog. Res.*, 1988, **3**, 125–130.
119. S. Cogoi, M. Paramasivam, B. Spolaore and L. E. Xodo, Structural polymorphism within a regulatory element of the human KRAS promoter: formation of G4-DNA recognized by nuclear proteins, *Nucleic Acids Res.*, 2008, **36**, 3765–3780.
120. S. Cogoi and L. E. Xodo, G-quadruplex formation within the promoter of the KRAS proto-oncogene and its effect on transcription, *Nucleic Acids Res.*, 2006, **34**, 2536–2549.
121. M. Paramasivam, S. Cogoi and L. E. Xodo, Primer extension reactions as a tool to uncover folding motifs within complex G-rich sequences: analysis of the human KRAS NHE, *Chem. Commun.*, 2011, **47**, 4965–4967.
122. M. Paramasivam, A. Membrino, S. Cogoi, H. Fukuda, H. Nakagama and L. E. Xodo, Protein hnRNP A1 and its derivative Up1 unfold quadruplex DNA in the human KRAS promoter: implications for transcription, *Nucleic Acids Res.*, 2009, **37**, 2841–2853.
123. R. K. Morgan, H. Batra, V. C. Gaerig, C. F. Hockings and T. A. Brooks, Identification and characterization of a new G-quadruplex forming

region within the kRAS promoter as a transcriptional regulator, *Biophys. Biochem. Acta, Gene Regul. Mech.*, 2015, **1859**(2), 235–245.
124. S. Cogoi, M. Paramasivam, V. Filichev, I. Geci, E. B. Pedersen and L. E. Xodo, Identification of a new G-quadruplex motif in the KRAS promoter and design of pyrene-modified G4-decoys with antiproliferative activity in pancreatic cancer cells, *J. Med. Chem.*, 2009, **52**, 564–568.
125. S. Cogoi, S. Zorzet, V. Rapozzi, I. Geci, E. B. Pedersen and L. E. Xodo, MAZ-binding G4-decoy with locked nucleic acid and twisted intercalating nucleic acid modifications suppresses KRAS in pancreatic cancer cells and delays tumor growth in mice, *Nucleic Acids Res.*, 2013, **41**, 4049–4064.
126. H. Brito, A. C. Martins, J. Lavrado, E. Mendes, A. P. Francisco, S. A. Santos, S. A. Ohnmacht, N. S. Kim, C. M. Rodrigues, R. Moreira, S. Neidle, P. M. Borralho and A. Paulo, Targeting KRAS Oncogene in Colon Cancer Cells with 7-Carboxylate Indolo[3,2-b]quinoline Tri-Alkylamine Derivatives, *PLoS One*, 2015, **10**, e0126891.
127. S. A. Ohnmacht, C. Marchetti, M. Gunaratnam, R. J. Besser, S. M. Haider, G. Di Vita, H. L. Lowe, M. Mellinas-Gomez, S. Diocou, M. Robson, J. Sponer, B. Islam, R. B. Pedley, J. A. Hartley and S. Neidle, A G-quadruplex-binding compound showing anti-tumour activity in an in vivo model for pancreatic cancer, *Sci. Rep.*, 2015, **5**, 11385.
128. A. Membrino, M. Paramasivam, S. Cogoi, J. Alzeer, N. W. Luedtke and L. E. Xodo, Cellular uptake and binding of guanidine-modified phthalocyanines to KRAS/HRAS G-quadruplexes, *Chem. Commun.*, 2010, **46**, 625–627.
129. C. P. Dinney, D. J. McConkey, R. E. Millikan, X. Wu, M. Bar-Eli, L. Adam, A. M. Kamat, A. O. Siefker-Radtke, T. Tuziak, A. L. Sabichi, H. B. Grossman, W. F. Benedict and B. Czerniak, Focus on bladder cancer, *Cancer Cell*, 2004, **6**, 111–116.
130. S. Cogoi, A. E. Shchekotikhin, A. Membrino, Y. B. Sinkevich and L. E. Xodo, Guanidino anthrathiophenediones as G-quadruplex binders: uptake, intracellular localization, and anti-Harvey-Ras gene activity in bladder cancer cells, *J. Med. Chem.*, 2013, **56**, 2764–2778.
131. S. Cogoi, A. E. Shchekotikhin and L. E. Xodo, HRAS is silenced by two neighboring G-quadruplexes and activated by MAZ, a zinc-finger transcription factor with DNA unfolding property, *Nucleic Acids Res.*, 2014, **42**, 8379–8388.
132. S. Cogoi, S. Zorzet, A. E. Shchekotikhin and L. E. Xodo, Potent Apoptotic Response Induced by Chloroacetamidine Anthrathiophenediones in Bladder Cancer Cells, *J. Med. Chem.*, 2015, **58**, 5476–5485.
133. S. Kumari, A. Bugaut and S. Balasubramanian, Position and stability are determining factors for translation repression by an RNA G-quadruplex-forming sequence within the 5′ UTR of the NRAS proto-oncogene, *Biochemistry*, 2008, **47**, 12664–12669.

134. S. Kumari, A. Bugaut, J. L. Huppert and S. Balasubramanian, An RNA G-quadruplex in the 5′ UTR of the NRAS proto-oncogene modulates translation, *Nat. Chem. Biol.*, 2007, **3**, 218–221.
135. A. Bugaut, R. Rodriguez, S. Kumari, S. T. Hsu and S. Balasubramanian, Small molecule-mediated inhibition of translation by targeting a native RNA G-quadruplex, *Org. Biomol. Chem.*, 2010, **8**, 2771–2776.
136. T. S. Dexheimer, D. Sun and L. H. Hurley, Deconvoluting the structural and drug-recognition complexity of the G-quadruplex-forming region upstream of the bcl-2 P1 promoter, *J. Am. Chem. Soc.*, 2006, **128**, 5404–5415.
137. M. I. Onyshchenko, T. I. Gaynutdinov, E. A. Englund, D. H. Appella, R. D. Neumann and I. G. Panyutin, Stabilization of G-quadruplex in the BCL2 promoter region in double-stranded DNA by invading short PNAs, *Nucleic Acids Res.*, 2009, **37**, 7570–7580.
138. H. Sun, J. Xiang, Y. Shi, Q. Yang, A. Guan, Q. Li, L. Yu, Q. Shang, H. Zhang, Y. Tang and G. Xu, A newly identified G-quadruplex as a potential target regulating Bcl-2 expression, *Biochim. Biophys. Acta*, 2014, **1840**, 3052–3057.
139. X.-D. Wang, T.-M. Ou, Y.-J. Lu, Z. Li, Z. Xu, C. Xi, J.-H. Tan, S.-L. Huang, L.-K. An, D. Li, L.-Q. Gu and Z.-S. Huang, Turning off transcription of the *bcl-2* gene by stabilizing the *bcl-2* promoter quadruplex with quindoline derivatives, *J. Med. Chem.*, 2010, **53**, 4390–4398.
140. S. Kendrick, H. J. Kang, M. P. Alam, M. M. Madathil, P. Agrawal, V. Gokhale, D. Yang, S. M. Hecht and L. H. Hurley, The dynamic character of the BCL2 promoter i-motif provides a mechanism for modulation of gene expression by compounds that bind selectively to the alternative DNA hairpin structure, *J. Am. Chem. Soc.*, 2014, **136**, 4161–4171.
141. B. Liu, H. Shang and D. Li, General peroxidase activity of a parallel G-quadruplex-hemin DNAzyme formed by Pu39WT – a mixed G-quadruplex forming sequence in the Bcl-2 P1 promoter, *Chem. Cent. J.*, 2014, **8**, 43.
142. R. Ferreira, R. Artali, J. Farrera-Sinfreu, F. Albericio, M. Royo, R. Eritja and S. Mazzini, Acridine and quindoline oligomers linked through a 4-aminoproline backbone prefer G-quadruplex structures, *Biochim. Biophys. Acta*, 2011, **1810**, 769–776.
143. Y. Long, Z. Li, J. H. Tan, T. M. Ou, D. Li, L. Q. Gu and Z. S. Huang, Benzofuroquinoline derivatives had remarkable improvement of their selectivity for telomeric G-quadruplex DNA over duplex DNA upon introduction of peptidyl group, *Bioconjug. Chem.*, 2012, **23**, 1821–1831.
144. Y. Wu, L. P. Zan, X. D. Wang, Y. J. Lu, T. M. Ou, J. Lin, Z. S. Huang and L. Q. Gu, Stabilization of VEGF G-quadruplex and inhibition of angiogenesis by quindoline derivatives, *Biochim. Biophys. Acta*, 2014, **1840**, 2970–2977.
145. J. Zhang, Q. Yu, Q. Li, L. Yang, L. Chen, Y. Zhou and J. Liu, A ruthenium(II) complex capable of inducing and stabilizing bcl-2 G-quadruplex

formation as a potential cancer inhibitor, *J. Inorg. Biochem.*, 2014, **134**, 1–11.
146. M. I. Onyshchenko, T. I. Gaynutdinov, E. A. Englund, D. H. Appella, R. D. Neumann and I. G. Panyutin, Quadruplex formation is necessary for stable PNA invasion into duplex DNA of BCL2 promoter region, *Nucleic Acids Res.*, 2011, **39**, 7114–7123.
147. R. Shahid, A. Bugaut and S. Balasubramanian, The BCL-2 5′ untranslated region contains an RNA G-quadruplex-forming motif that modulates protein expression, *Biochemistry*, 2010, **49**, 8300–8306.
148. R. De Armond, S. Wood, D. Sun, L. H. Hurley and S. W. Ebbinghaus, Evidence for the presence of a guanine quadruplex forming region within a polypurine tract of the hypoxia inducible factor 1α promoter, *Biochemistry*, 2005, **44**, 16341–16350.
149. Y. Qin, E. M. Rezler, V. Gokhale, D. Sun and L. H. Hurley, Characterization of the G-quadruplexes in the duplex nuclease hypersensitive element of the PDGF-A promoter and modulation of PDGF-A promoter activity by TMPyP4, *Nucleic Acids Res.*, 2007, **35**, 7698–7713.
150. E. Salvati, P. Zizza, A. Rizzo, S. Iachettini, C. Cingolani, C. D'Angelo, M. Porru, A. Randazzo, B. Pagano, E. Novellino, M. E. Pisanu, A. Stoppacciaro, F. Spinella, A. Bagnato, E. Gilson, C. Leonetti and A. Biroccio, Evidence for G-quadruplex in the promoter of vegfr-2 and its targeting to inhibit tumor angiogenesis, *Nucleic Acids Res.*, 2014, **42**, 2945–2957.
151. D. Sun, K. Guo and Y. J. Shin, Evidence of the formation of G-quadruplex structures in the promoter region of the human vascular endothelial growth factor gene, *Nucleic Acids Res.*, 2011, **39**, 1256–1265.
152. M. T. Malik, M. G. O'Toole, L. K. Casson, S. D. Thomas, G. T. Bardi, E. M. Reyes-Reyes, C. K. Ng, K. A. Kang and P. J. Bates, AS1411-conjugated gold nanospheres and their potential for breast cancer therapy, *Oncotarget*, 2015, **6**, 22270–22281.
153. E. Y. Lam, D. Beraldi, D. Tannahill and S. Balasubramanian, G-quadruplex structures are stable and detectable in human genomic DNA, *Nat. Commun.*, 2013, **4**, 1796.
154. B. J. Wright, N. J. Galtieri and M. Fell, Non-adherence to prescribed home rehabilitation exercises for musculoskeletal injuries: the role of the patient-practitioner relationship, *J. Rehabil. Med.*, 2014, **46**, 153–158.
155. P. M. Yangyuoru, M. Di Antonio, C. Ghimire, G. Biffi, S. Balasubramanian and H. Mao, Dual binding of an antibody and a small molecule increases the stability of TERRA G-quadruplex, *Angew. Chem.*, 2015, **54**, 910–913.
156. R. V. Brown, V. C. Gaerig, T. Simmons and T. A. Brooks, Helping Eve Overcome ADAM: G-Quadruplexes in the ADAM-15 Promoter as New Molecular Targets for Breast Cancer Therapeutics, *Molecules*, 2013, **18**, 15019–15034.
157. S. T. Hsu, P. Varnai, A. Bugaut, A. P. Reszka, S. Neidle, S. Balasubramanian and A. G-rich, sequence within the c-kit oncogene

promoter forms a parallel G-quadruplex having asymmetric G-tetrad dynamics, *J. Am. Chem. Soc.*, 2009, **131**, 13399–13409.
158. A. T. Phan, V. Kuryavyi, S. Burge, S. Neidle and D. J. Patel, Structure of an unprecedented G-quadruplex scaffold in the human *c-kit* promoter, *J. Am. Chem. Soc.*, 2007, **129**, 4386–4392.
159. S. Rankin, A. P. Reszka, J. Huppert, M. Zloh, G. N. Parkinson, A. K. Todd, S. Ladame, S. Balasubramanian and S. Neidle, Putative DNA quadruplex formation within the human c-kit oncogene, *J. Am. Chem. Soc.*, 2005, **127**, 10584–10589.
160. A. K. Todd, S. M. Haider, G. N. Parkinson and S. Neidle, Sequence occurrence and structural uniqueness of a G-quadruplex in the human c-kit promoter, *Nucleic Acids Res.*, 2007, **35**, 5799–5808.
161. S. Bonnal, C. Schaeffer, L. Creancier, S. Clamens, H. Moine, A. C. Prats and S. Vagner, A single internal ribosome entry site containing a G quartet RNA structure drives fibroblast growth factor 2 gene expression at four alternative translation initiation codons, *J. Biol. Chem.*, 2003, **278**, 39330–39336.
162. A. Ambrus, D. Chen, J. Dai, T. Bialis, R. A. Jones and D. Yang, Human telomeric sequence forms a hybrid-type intramolecular G-quadruplex structure with mixed parallel/antiparallel strands in potassium solution, *Nucleic Acids Res.*, 2006, **34**, 2723–2735.
163. J. Li, J. J. Correia, L. Wang, J. O. Trent and J. B. Chaires, Not so crystal clear: the structure of the human telomere G-quadruplex in solution differs from that present in a crystal, *Nucleic Acids Res.*, 2005, **33**, 4649–4659.
164. S. L. Palumbo, S. W. Ebbinghaus and L. H. Hurley, Formation of a unique end-to-end stacked pair of G-quadruplexes in the hTERT core promoter with implications for inhibition of telomerase by G-quadruplex-interactive ligands, *J. Am. Chem. Soc.*, 2009, **131**(31), 10878–10891.
165. A. M. Zahler, J. R. Williamson, T. R. Cech and D. M. Prescott, Inhibition of telomerase by G-quartet DNA structures, *Nature*, 1991, **350**, 718–720.
166. P. Agarwala, S. Pandey, K. Mapa and S. Maiti, The G-quadruplex augments translation in the 5′ untranslated region of transforming growth factorbeta2, *Biochemistry*, 2013, **52**, 1528–1538.
167. Y. Xu and H. Sugiyama, Formation of the G-quadruplex and i-motif structures in retinoblastoma susceptibility genes (Rb), *Nucleic Acids Res.*, 2006, **34**, 949–954.
168. W. Huang, P. J. Smaldino, Q. Zhang, L. D. Miller, P. Cao, K. Stadelman, M. Wan, B. Giri, M. Lei, Y. Nagamine, J. P. Vaughn, S. A. Akman and G. Sui, Yin Yang 1 contains G-quadruplex structures in its promoter and 5′-UTR and its expression is modulated by G4 resolvase 1, *Nucleic Acids Res.*, 2012, **40**, 1033–1049.

Subject Index